Perspectives on Classification in Synthetic Sciences

This volume launches a new series of contemporary conversations about scientific classification. Most philosophical conversations about kinds have focused centrally or solely on natural kinds, that is, kinds whose existence is not dependent on the scientific process of synthesis. This volume refocuses conversations about classification on unnatural, or synthetic, kinds via extensive study of three paradigm cases of unnatural kinds: nanomaterials, stem cells, and synthetic biology.

Julia R. S. Bursten is Assistant Professor in the Department of Philosophy at The University of Kentucky, USA.

History and Philosophy of Technoscience
Series Editor: Alfred Nordmann

Titles in this series:

Environments of Intelligence
From Natural Information to Artificial Interaction
Hajo Greif

A History of Technoscience
Erasing the Boundaries between Science and Technology
David F. Channell

Visual Representations in Science
Concept and Epistemology
Nicola Mößner

From Models to Simulations
Franck Varenne

The Reform of the International System of Units (SI)
Edited by Nadine de Courtenay, Olivier Darrigol, and Oliver Schlaudt

The Past, Present, and Future of Integrated History of Philosophy of Science
Edited by Emily Herring, Konstantin S. Kiprijanov, Kevin Jones and Laura M. Sellers

Nanotechnology and Its Governance
Arie Rip

Perspectives on Classification in Synthetic Sciences
Unnatural Kinds
Edited by Julia R. S. Bursten

For more information about this series, please visit: www.routledge.com/History-and-Philosophy-of-Technoscience/book-series/TECHNO

Perspectives on Classification in Synthetic Sciences

Unnatural Kinds

Edited by Julia R. S. Bursten

Routledge
Taylor & Francis Group

LONDON AND NEW YORK

First published 2020
by Routledge
2 Park Square, Milton Park, Abingdon, Oxon OX14 4RN

and by Routledge
605 Third Avenue, New York, NY 10017

First issued in paperback 2021

Routledge is an imprint of the Taylor & Francis Group, an informa business

British Library Cataloguing-in-Publication Data
A catalogue record for this book is available from the British Library

Library of Congress Cataloging-in-Publication Data
Names: Bursten, Julia R. S., editor.
Title: Perspectives on classification in synthetic sciences : unnatural kinds / edited by Julia R. S. Bursten.
Description: Milton Park, Abingdon, Oxon ; New York, NY : Routledge, 2020. | Series: History and philosophy of technoscience | Includes bibliographical references and index.
Identifiers: LCCN 2019011601 (print) | LCCN 2019014185 (ebook) | ISBN 9781351581301 (adobe) | ISBN 9781351581288 (mobi) | ISBN 9781351581295 (epub) | ISBN 9781138298101 (hbk) | ISBN 9781315098838 (ebk)
Subjects: LCSH: Synthetic biology—Classification. | Bioengineering. | Nanotechnology.
Classification: LCC TA164 (ebook) | LCC TA164 .P47 2020 (print) | DDC 660.6—dc23
LC record available at https://lccn.loc.gov/2019011601

ISBN 13: 978-0-367-78589-5 (pbk)
ISBN 13: 978-1-138-29810-1 (hbk)

Typeset in Times New Roman
by Apex CoVantage, LLC

Contents

Figures and Tables

Figures

Tables

Contributors

Bryan A. Bartley is a Scientist at Raytheon BBN Technologies in Cambridge, MA. His research involves knowledge representation and data integration in systems and synthetic biology. He received his PhD in Bioengineering from the University of Washington, Seattle.

Julia R. S. Bursten is an Assistant Professor of Philosophy at the University of Kentucky and former co-chair of the Philosophy of Science Association Women's Caucus. Her research concerns questions of modeling and scale in nanoscience and other physical sciences. From 2011 to 2015, she served as the resident philosopher in the Millstone Nanosynthesis Laboratory at the University of Pittsburgh.

Melinda Bonnie Fagan is an Associate Professor of Philosophy at the University of Utah, where she holds the Sterling M. McMurrin Chair. Her research focuses on experimental practice in biology (particularly stem cell and developmental biology), explanation, and modeling. She is the author of *Philosophy of Stem Cell Biology: Knowledge in Flesh and Blood* (Palgrave Macmillan, 2013) and more than 40 articles and book chapters on topics in philosophy of science and biology.

Evan Hepler-Smith is an Assistant Professor of History at Duke University. His research addresses the interconnected development of global chemical industries, information technologies, and environmental toxicity since the mid-nineteenth century.

Catherine Kendig is an Assistant Professor of Philosophy at Michigan State University. She completed her PhD in Philosophy at the University of Exeter/ESRC Centre for Genomics in Society and her MSc in Philosophy and History of Science at King's College London. Her main research interests are in philosophy of scientific classification, natural kinds, synthetic biology, socially engaged philosophy of science, and philosophy of race. Her research in the metaphysics of synthetic biology has been supported through the National Science Foundation, Division of Molecular and Cellular Biosciences. She is editor of the recent collection of interdisciplinary essays, *Natural Kinds and Classification in Scientific Practice* (Routledge, 2016).

Vadim Keyser is an Assistant Professor in the Philosophy Department as well as affiliated faculty in the MS Biotechnology Program at California State University, Fresno. His research focuses on the complex relationships between measurement, experimentation, and technology in the physical and biological sciences.

Jill E. Millstone is an Associate Professor of Chemistry at the University of Pittsburgh with affiliated appointments in the Departments of Chemical Engineering and Mechanical Engineering and Materials Science. She has received honors including the NSF CAREER Award, the ACS Unilever Award, the Cottrell Research Scholar Award, and the Kavli Foundation Emerging Leader in Chemistry lectureship. Currently, she serves as an Associate Editor at ACS Nano and on the editorial advisory boards of both *Nanoscale* and *Journal of Physical Chemistry Letters*. She is also the chair of the Nanoscience Division within the Inorganic Division of the American Chemical Society. Her group studies the chemical mechanisms underpinning metal nanoparticle synthesis, surface chemistry, and optoelectronic behaviors.

Ubaka Ogbogu is an Associate Professor in the Faculty of Law at the University of Alberta. He holds a cross-appointment in the Faculty of Pharmacy and Pharmaceutical Sciences. He is a recipient of the Confederation of Alberta Faculty Associations Distinguished Academic Early Career Award and has been nominated for the Faculty of Law's Tevie Miller Teaching Award. His research examines the philosophical, legal, and ethical questions arising from or associated with cutting-edge biotechnologies and the role of law in mediating encounters between science and society.

Aleta Quinn is an Assistant Professor of Philosophy at the University of Idaho and a Research Collaborator at the Smithsonian Institution's National Museum of Natural History. She works on the history and philosophy of biology with a special focus on phylogenetic inference and species delimitation. She is an avid proponent of citizen scientists, contributing to several range and natural history studies as well as providing hundreds of records to iNaturalist.org.

John Rumble, Jr., is President and CEO of R&R Data Services of Gaithersburg, MD. He has long been active in materials informatics and was the Chair of the CODATA Working Group on Nanomaterials. He was Director of the NIST Standard Reference Data Program from 1994 to 2000 and was awarded the 2006 CODATA Prize.

1 Introduction

Julia R. S. Bursten

This volume began in a series of enthusiastic conversations between myself and Evan Hepler-Smith, one of this volume's contributors, about an observed oversight in both the historical and the philosophical literature on classification. Hepler-Smith, a historian of chemistry, and I, a philosopher whose work often draws on chemistry, found common ground in a shared frustration with our disciplines' emphases on the chemical elements as the stereotypical example of a natural kind. The frustration we shared was that while the elements did display many hallmarks of paradigmatic kindhood, elements were not the kinds of kinds that generated interesting challenges for classification in chemistry, nor even were they the kinds of kinds that occupied much contemporary critical chemical thought. Compounds, complexes, reaction pathways, substrates, solutions – these were the kinds of the chemistry laboratory, and rarely if ever did they slot neatly into taxonomies in the orderly manner of classification suggested by the Periodic Table of Elements. A focus on the rational and historical basis of the development of the Periodic Table had made the received view of chemical classification appear far more pristine, and far less interesting, than either of us believed it to be.

From this observation, we sought out other examples of kinds that behaved in unusual or unexpected ways given their disciplinary origins, and we began to notice a commonality among many of the kinds of kinds we encountered: they were made, synthesized, grown, or otherwise brought to the attention of scientific investigation not by mere detection, but by some active construction on the part of investigators. While many researchers in other traditions represented in this series, notably those working in Science and Technology Studies, have investigated classification specifically in the context of synthetic kinds, we believed the literature in the philosophy of science was lagging.

From natural to unnatural kinds

Conversations about classification and categorization in philosophy are as old as the discipline itself, and philosophical problems about classification are as pervasive as they are persistent. At their roots, the problems of classification are those of how things are individuated from other things, and how things are connected to their properties. On some level, the problem of classification is no less than the aim of philosophy itself, as infamously articulated by Wilfrid Sellars: how do things, in the broadest possible sense of the term, hang together, in the broadest possible sense of the term?

Given the broad scope of these problems, it is unsurprising that the problems of classification have ramifications in metaphysics, epistemology, philosophy of language, philosophy of science, and even logic. The connection between things and their properties can be properly understood, rationalized, and theorized about only when it is known what the things are, what the properties are, what the rules of connection are, and what parts of worldly experience generate those rules: essences, microstructural or ineffable; semantic conventions; psychological or inferential habits; or something else entirely. Canonical philosophical works in classification and individuation all aim to address questions about how the things out there in the world came to be connected to their properties in the ways they are connected, and how those thing-property connections can make sense of the behavior of words, objects, and the people who use them.

This volume is different. Our goal in the essays here is not to address the connections between things and their properties in nature. While all of the contributions to this volume are thoroughly naturalistic, the aim of this collection is not to contemplate the joints at which nature might be carved. Rather, the authors in this volume all came together around a question that centuries of philosophical writing on classification seems to have left unanswered: how does the connection between things and their properties change when part of that connection consists in the intentional seeking and joining of a cluster of properties that have never before been clustered in just such a way? How, in other words, does classification work when the kinds in question are not natural, but synthetic?

The kinds of kinds considered in this volume, which we describe variously as "unnatural" or "synthetic" kinds, fall under a certain category of experimental products found most often in chemistry and biology, but which may be found across the sciences. These kinds are often, although not exclusively, the product of an experimental process identified as synthesis. Many synthetic kinds are the products of synthesis, and many syntheses produce synthetic kinds, but being a product of synthesis is neither necessary nor sufficient for being a synthetic kind. Some synthetic things are not synthetic kinds of the sort we address here, and in order to develop a picture of the synthetic kinds that will be the focus of the essays that follow, it will be useful to begin by sketching the negative.

To this end, consider salicylic acid, which is the compound responsible for the pain-relieving properties of willow bark and a precursor to the active ingredient in aspirin, acetylsalicylic acid. Salicylic acid is a metabolic by-product of the glucoside salicin. When salicin is introduced into the digestive systems of certain life forms, including humans, the compound is broken down into glucose and salicyl alcohol. Salicyl alcohol is then oxidized to form salicylic acid. However, it is also possible to produce salicylic acid in a laboratory, either by mimicking in beakers the natural digestive reaction mechanism described earlier or, more commonly, through a mechanism that combines phenol with sodium hydroxide, carbon dioxide, and sulfuric acid. This latter synthetic process is distinct from the former in that it does not imitate a "natural" process; nonetheless, the product it produces is chemically identical to the derivative of salicin. So while salicylic acid produced by the phenol mechanism may appropriately be called "synthetic salicylic acid," neither "salicylic acid" nor "synthetic salicylic acid" are examples of synthetic kinds, because the two acids are chemically identical and because the structure is one that is, for lack of a better term, naturally occurring. The phenol mechanism is synthetic, in contrast with the digestive mechanism, but the product is not a synthetic kind of the sort that concerns the authors in this collection.

Synthetic kinds are not replications of natural categories, even ones produced by radical construction techniques. Rather, they are definitively unnatural kinds, products of a targeted search for groups of properties that do not regularly group together: genetically enhanced kittens that glow in the dark, clusters of carbon atoms a few billionths of a meter wide that form the wheels and axles of a car and roll around, plutonium, and most plastics. Synthetic kinds adapt, evolve, re-engineer, or otherwise borrow pieces or properties from known, and often natural, things in order to make something new.

Synthetic kinds encounter many of the same and quite a few distinct challenges, relative to their non-synthetic counterparts. The most obvious difference is that in the case of synthetic kinds, a classification system is not expected to carve nature at its joints, so there is no (pardon the pun) natural tendency to connect conclusions about classification to a more robust metaphysical theory. This is not to say that there are not interesting ontological questions or implications to be drawn from the study of synthetic kinds. Instead, the point is just that where many theories of natural kinds are motivated by a desire to discover the grammar of the book of nature, no such grounding underlies the investigation of synthetic kinds.

A practical upshot of this difference, which was noticed by a few authors in this volume, is that the philosophical preference for taxonomies is less well-motivated in the study of synthetic kinds than in the study of natural kinds. Taxonomies are classification systems without cross-cutting divisions, that is, ones where every subclass is a proper part of exactly one class. The category "quadruped" cross-cuts the standard species classes of mammals and reptiles, since some mammals and some reptiles are quadrupedal (dogs and salamanders), and some are non-quadrupedal (humans and snakes). The tendency of theories of natural kinds to be employed in the articulation of essences, and the tendency of philosophers to identify essences as singular and unique to the objects of which they are the essence, has led to a preference for theories of natural kinds that produce unique and singular, and so non-cross-cutting, classification systems (although whether that preference may be realized is quite another matter entirely). The grounds for these tendencies fall away in discussions of synthetic kinds, and in this collection, the utility – and occasionally, the necessity – of cross-cutting categories is treated not as a bug but as a feature of many synthetic classification systems.

The role of human use in the generation, articulation, and classification of synthetic kinds is also somewhat distinct from both realist and some constructivist perspectives on the role of human use in the articulation of natural kinds. In the philosophical literature on natural kinds, the realist position holds that kinds are in some way connected to mind-independent groupings in nature, whether those groupings are unified on the basis of an essence, a cluster of properties, or something else. Constructivist positions, on the other hand, typically assert that kind categories are the product of contingent human activity, that the kinds in use are a reflection not of the contours of the world but of the incidental history of human projects so far. The latter position is certainly consistent with what has been said so far about synthetic kinds, and certain versions of the former could be as well – for instance, a view that as-yet uncreated synthetic kinds are non-actual possibles generated by the natural world, coupled with realism about non-actual possibles. However, the motivation for resolving this dispute one way or another drops out of many of the discussions about synthetic kinds that make up the following pages, as it has for other recent philosophical work on kinds that, like this volume does, subscribes to the practice turn in philosophy of science.

It is inarguable that both the scientific objects and the classificatory groupings in this volume are constructed. The groupings may also be real, in some sense that is analogous to the stuff of nature carved at its joints. The problem is that settling whether or not synthetic kinds behave like natural kinds does not advance most of the philosophical inquiries considered by the authors of this volume, such as how to understand the role of a stem cell line's origin as part of its classificatory scheme, or whether information about a nanomaterial solution's solvent is information about the nanomaterial or about its environment. A central motivation for many of the authors in this volume has been to begin the philosophical conversation about classification from these alternative trailheads, and to see whether the path returns us to the well-worn tracks in the metaphysics and semantics of kinds. By and large, the contributions in this volume diverge from those roads, and it is our hope that by blazing new trails, we can lay the groundwork for a new set of conversations about the philosophical foundations of classification, moving even beyond the pragmatic approaches of latter-day philosophy of science and into collaborative, interdisciplinary considerations of the limits, uses, and logical and practical underpinnings of scientific classification.

How to read this collection

This volume grew out of a workshop, "Unnatural Kinds: Interdisciplinary Perspectives on Scientific Classification," held in 2016 at San Francisco State University and organized by Hepler-Smith and myself. As a result of this collaborative beginning, many of the pieces in the volume draw on each other, both directly and indirectly. This has impacted the structure of the collection in two ways. First, where appropriate, the authors and I have indicated specific sites of internal cross-reference between individual pieces in the volume, which are noted by parenthetical reference to the cross-reference's chapter number (e.g., "See Chapter 3."). Second, the collection has been organized into four thematic sections, each containing only two essays, which I have called "Dialogues." This organization is meant to encourage readers to view the pieces in each section as in conversation with one another. It is intended to recapture some of the spirit of interdisciplinary discourse that underwrote the workshop where this project began, as well as to provide a series of tractable inroads to the themes of the volume from a variety of angles. Each Dialogue, like each essay, may be read individually, but the hope is that together the pieces of the collection themselves generate a synthesis, a conversation whose whole transcends the sum of its parts.

Dialogue One: Historical Lenses on Classification in Chemistry and Biology comprises two essays that study historical instances of classification from the latter days of the 19th century and the early part of the 20th. The authors differ in both the cases they treat and the historiographic approaches they take, and together these pieces set the stage for the contemporary cases that follow by providing a complex and multifaceted picture of the aims and considerations that underwrote scientific decisions about classification just over a century ago.

The Dialogue begins with Evan Hepler-Smith's "Crafting names and making kinds: lessons from the 1892 Geneva Nomenclature Congress." Hepler-Smith's careful study of the 1892 congress on naming practices for organic compounds unpacks a series of practical and philosophical considerations that contributed to a new conception of the connection between nomenclature and classification. Today, organic nomenclature is often held up as an example of the power of clean taxonomy: the nomenclature system decomposes compounds into functional groups and structural foundations, each of which form the basis for useful classificatory work in the laboratory. The ideal of a system like that of organic nomenclature is often referenced with respect to classification in other areas of synthetic classification, such as nanoscience, as evidenced in the essays of Dialogue Four. Hepler-Smith's account of the origins of this system reveal a concerted and often fraught effort, whose story should inform any contemporary view of classification that employ ideas of structural decomposability – as most contemporary classification schemes in chemistry and biology do. The historical lessons from Hepler-Smith's account add texture in particular to the accounts of classification via structural decomposition developed in the contributions of Kendig and Bartley, Rumble, Fagan, and Millstone and Bursten.

Hepler-Smith's account is followed by Aleta Quinn's "Biological kinds at the turn of the 20th century: characters, genes, and species as theoretical elements." Quinn begins with a rich intellectual history of the philosophical perspectives on classification that influenced scientific views during the period she and Hepler-Smith both consider. Beginning with John Stuart Mill, Quinn deftly unpacks a view of the stakes of conversations about natural kinds for philosophers, biologists, and chemists at the end of the 19th century. Her account begins in an oft-overlooked era of work in natural kinds, when philosophers and chemists used biological species as a stereotype or paradigm on which to base discussions of chemical elements, rather than the other way around. The chapter focuses on William Bateson's "theoretical elements" view of species kinds, which, Quinn argues, encounters the analogy between chemical and biological kinds at an inflection point when it begins to switch to the view more common today, namely that chemical elements are a paradigmatic kind toward whose standards species taxonomies should strive. This investigation of the comparison between biological and chemical kinds in the construction of early 20th-century theories of heredity serves as a complement to the chemical story of the Nomenclature Congress, as well as a historical and philosophical foundation for the remaining Dialogues' continued engagement with the tangled classificatory relations between chemistry and biology.

Dialogue Two: A New Synthesis of Concerns about Biological Kinds considers a variety of contemporary biological kinds, each of which poses a set of challenges quite unlike those encountered in conversations about speciation and heredity. Vadim Keyser begins this Dialogue with "Artifacts and artefacts: a methodological classification of context-specific regularities," in which he constructs a classificatory scheme for biological objects that are neither natural kinds nor mere artifacts. Through case studies on arsenic-consuming living organisms and the measurement of ecological effects on epigenetics in reptilian embryos, Keyser argues that these context-sensitive experimental kinds, which he calls "artefacts," serve important and distinctive roles in the advancement of biological experiment and theory, and as such, diverge from mere artifacts.

The second part of this Dialogue considers a problem for classification in biology that is removed even further from standard accounts of natural kinds. Catherine Kendig and Bryan A. Bartley's "Synthetic kinds: kind-making in biology" is a focused and interdisciplinary interrogation of the notion of a biological part in synthetic biology, a branch of the field concerned with the construction from biological parts of living systems. The foundation of synthetic biology rests on the combination and recombination of parts, and Kendig (a philosopher) and Bartley (a bioengineer) investigate both the philosophical challenges for and biologists' practical solutions to the problem of part identification and individuation in synthetic biology. They chronicle the development of biological part repositories, such as the Synthetic Biology Open Language, and discuss how the multivariate schemes in parts repositories address the challenges to part naming and tracking that are posed by synthetic biology's disciplinary backgrounds in both structural biology and engineering.

Kendig and Bartley's chapter stands here in dialogue with Keyser's chapter due both to their shared biological subject matter and their shared motivation concerning the need for philosophical accounts to reflect biological practice, although the two pieces develop different facets of a view on classification of synthetic biological kinds. However, readers interested in the role of programmable databases and computer ontologies in the classification of synthetic kinds will find it instructive to read Kendig and Bartley's contribution as a complement to John Rumble, Jr.'s discussion of nanomaterials ontologies in Dialogue Four.

Dialogue Three: Scientific, Philosophical, and Legal Challenges in Classifying Biological Constructs offers a variety of perspectives on the classificatory concerns raised by the development of stem cells as a research object. Legal scholar Ubaka Ogbogu begins the Dialogue with "What is a new object? Case studies of classification problems and practices at the intersection of law and biotechnology." This riveting discussion grounds itself in the practical recognition that for certain types of controversial synthetic kinds in biology, such as clones, human embryonic stem cells, and genetically modified organisms, the need for fine-grained scientific determinations regarding classification is often motivated by extrascientific concerns, such as patentability and legally guaranteed protections for the sanctity of life. Ogbogu's central case studies treat the question of whether cloned organisms count as embryos, the ontological status of excised human tissue, and the patentability of transgenic mice, but the chapter also serves as a targeted primer on the legal history of synthetic-kind classification in biology in the United Kingdom, Canada, and the United States. As such, it makes a unique and important contribution to many of the discussions in the volume, although the cases it treats speak most directly to the account developed by Melinda Bonnie Fagan in the second part of Dialogue Three.

Fagan's chapter, "Stem cells and nanomaterials as experimental kinds," builds on her work establishing the philosophy of stem cell biology and develops her account of a stem cell as a particular sort of experimental kind. Fagan's account is based not on a particular philosophical approach to classification, but rather on a minimal abstract model of a stem cell, consisting of lineage, variable characters, time/generation intervals, and a developmental process of cell specialization. Her goals in this account are to describe the behaviors of stem cells as they are individuated and employed in scientific practice, to outline the challenge of classifying stem cells as individuals, and to construct a basis of comparison across synthetic sciences.

Once her classification scheme is established, Fagan compares it with the scheme I have developed in writing about nanomaterials, and which I discuss with Jill E. Millstone in our contribution to the volume. This comparative project sketches the beginning of a more general view of classification in synthetic sciences. It also provides a particularly useful entry point for readers with a biological background who wish to engage with the volume's chemical contributions. Moreover, the comparative project Fagan undertakes is in many ways at the heart of the volume: to discover the shared points of divergence between synthetic and natural kinds, and to use these points to reflect back on what a philosophical theory of scientific classification might look like if it could accommodate both natural and unnatural kinds. Fagan's direct confrontation of the comparison between stem cells and nanomaterials crystalizes quite a few of the themes in the earlier essays. Her careful consideration of the analogy between stem cell and nanomaterial kinds also echoes the historical questions of analogy raised in Quinn's contribution.

The closing Dialogue returns to chemistry and to chemical nanoscience. Unlike the previous Dialogues, Dialogue Four: Synthetic Kinds in Chemistry and Nanoscience, features a majority of practicing chemists, all of whom have spent time and energy engaging with philosophical considerations of classification in their own work. The contributions of these practitioners illustrate the practical need for a theory of classification that accommodates the synthetic sciences, and in particular the classificatory challenges imposed by the central chemical activity of synthesis itself. The Dialogue begins with a transcribed interview I conducted with Jill E. Millstone, a synthetic nanochemist with an active and lauded academic research profile. This contribution, "Nanochemistry meets philosophy of science: a conversation about collaborative classification," aims to capture not just the philosophical and scientific content but also the collaborative tenor of Millstone's and my work on classification in nanoscience. It introduces readers to the outlines of the account of nanomaterials classification I have developed elsewhere, and which owes an intellectual debt to Millstone's work and to our longtime collaboration. Moreover, the interview illustrates a chemist's perspective on the classificatory problems that plague nanoscience, pointing not just to a lack of individuation framework but to a deep conceptual challenge at the heart of nanoscience itself. As we realize over the course of the interview, questions of differentiating nanomaterials cannot be satisfactorily answered without addressing the challenge of how to conceive of nanosynthesis itself, and conceptions of synthesis from other areas of chemistry are not up to the task.

The next piece in the Dialogue is Rumble's "Categorization of nanomaterials: tools for being precise." Rumble, a chemist by training with a long and illustrious career in data science, has been a key figure in the development of a nanomaterials ontology for the Committee on Data for Science and Technology (CODATA) of the International Council for Science (now part of the International Science Council). His article motivates and articulates the CODATA framework and provides a useful contrast to the scheme discussed in my and Millstone's contribution. In particular, Rumble raises practical concerns about the need for standardization of nanoparticle safety tests, which engender a finer-grained classification scheme than the one Millstone and I propose. Rumble's CODATA scheme is also grounded firmly in the use of computer ontologies for classification, which raises its own set of challenges and advantages.

Shared prospects for a theory of synthetic kinds

Together, this collection of essays is meant to be read as the beginning of an interdisciplinary conversation about classification across the synthetic sciences, motivated not only by traditional philosophical concerns about the conceptual machinery of categorization, but also by a number of pressing and practical challenges imposed by inadequate classificatory schemes in both scientific and extra-scientific settings. The objects of science considered by the authors are the stuff of everyday science, and by interrogating these kinds, the authors in this collection generate an array of considerations that diverge from the usual philosophical conversations.

While our aim in assembling this collection was never to propose a unified theory of synthetic kinds, some common themes do emerge across the Dialogues, which may motivate and constrain such a theory. First and foremost, it is clear that an account of synthetic kinds will have to be informed by a detailed understanding of the role of classification in historical and contemporary scientific practice. Each of the pieces in this collection considers specific instances of kind use, using details of these case studies both to highlight commonalities and to demonstrate the limits of general claims about the nature of synthetic kinds. Further, quite a few of the pieces in this collection are written either by practitioners or former practitioners, or as a collaboration between practitioners and philosophers. The attention to practice is not a one-way street; many of our authors consult with and inform scientific practice with their investigations.

Second, as a consequence of the attention to scientific practice that generated the accounts in this collection, there is a strong thread of attention to the methodological or pragmatic role of kind categories in scientific settings. As I have argued in previous work, this role for kind categories is distinct from the metaphysical and semantic roles for kind categories in philosophical accounts of kinds. By "the methodological role of kind categories," I mean the uses to which kind categories are put in the carrying-out of scientific projects and projects that rely on scientific categories. (The "methodological" vocabulary is meant to evoke the idea that use of these categories is endemic to the exercise of scientific methods.) For example, Ogbogu's focused study of the role of particular synthetic-kind categories in establishing legal rulings provides a clear exemplar of how the use of these categories impacts both the lives of individuals and regulations on research and intellectual property. Further, the accounts of classification proposed or gestured at in the chapters by Keyser, Kendig and Bartley, and Fagan all reference methodological criteria for establishing kindhood. On the practitioners' side, too, the CODATA criteria that Rumble outlines include features of nanomaterials that relate to their method of synthesis.

Third, there is a tacit preference for fineness of grain suggested by a number of the discussions in the essays. This is not universal – Millstone and I, in particular, consider the limits of a narrowly divided classification system for nanomaterials – but in a number of the essays, the preference for contextual categorization generates a natural tendency toward "splitting" over "lumping," which seems driven in part by the non-naturalness of synthetic kinds. In Hepler-Smith's account of the naming conventions of organic chemistry, in Kendig and Bartley's biological parts repositories, and in the CODATA nanomaterials ontology, there are built-in mechanics (through name adjustments or adding a tag in an ontology) to denote that an object has a particular feature that distinguishes it from similar objects that lack that feature – the location of a functional group on the second vs. the third atom in a long-chain carbon, the notation of lineage in identifying a stem cell, or the tagging of a biological part in the iGEM registry with an engineering function (e.g., "chassis").

Moreover, devising systems with epistemic frameworks that support fine-grained splitting appears to be something of a necessity for classification in synthetic sciences, since part of the whole point of synthetic kinds is to generate new objects that do not yet conform to the categories encoded thus far. The fineness of grain is typically accompanied by an openness in the machinery of a synthetic-kind classification system, which allows for novel features to be added into a naming convention or ontology. This openness is usefully contrasted with the kind of openness in the Periodic Table of Elements, where well-known features suggest the possibility of as-yet undetected objects with defined characteristics. The Periodic Table is a closed machinery with open possible objects, where the naming conventions of organic chemistry are more open, in that they are at least combinatorially unbounded. Contemporary ontologies of synthetic kinds expand even further on this in-principle limitlessness, not only opening the possible combinations of names but also leaving explicit room in the system to identify previously unrecognized figures of merit, for instance by tagging objects in the ontology with tags that are devised well after the ontology's creation.

Taken together, these themes of (1) collaborative interaction with practice, (2) founding an account of synthetic kindhood on the methodological or pragmatic role of kinds, and (3) devising intentionally fine-grained systems with machinery that allows for inclusion of as-yet unrecognized figures of merit, blaze a trail that we hope future theories of synthetic classification will continue to help us clear. We are grateful to have had the opportunity to collect our thoughts on synthetic classification thus far, and we look forward to where these conversations will take us.

Acknowledgments

Many people and institutions assisted in making this volume possible, and in the spirit of fine-grained classification, I would like to acknowledge them individually. As discussed earlier, Evan Hepler-Smith and I conceived of the workshop that first catalyzed the project together, and his fingerprint marks many of the pages that follow. Funding for the workshop was provided by the San Francisco State University College of Liberal and Creative Arts and Department of Philosophy, and Jennifer Waller offered valuable organizational assistance. Evan Bolton, Will Fischer, Michael Hartmann, James Overton, and Alok Srivastava also presented research at that workshop, and while their projects do not constitute independent chapters in this collection, their influence is deeply felt throughout. In particular, Srivastava's theory of the making-history of a synthetic object has generated moments of epiphany in my own work, and the absence of his account in this collection is, in my view, the volume's greatest weakness. Aaron Chavez, Austin Due, Robert Kok, and Mitch Enriquez also provided both organizational assistance and scholarly contributions through their involvement in the workshop, and Elihu Gerson's participation as a particularly spirited discussant was welcome and insightful.

Alfred Nordmann, who edits the History and Philosophy of Technoscience series, encouraged the development of the workshop into the present volume, and without his buoying enthusiasm for this work the collection surely would not have materialized. With the support of Technische Universität Darmstadt and the CompuGene project, Alfred has been instrumental, as well, in providing opportunities for some of the authors to come together over the course of the project and conduct further research together. I am also grateful for the tireless support of the editorial assistants at Routledge, Michael Bourne, Julie Fitzsimmons, and Dana Moss, and for the deeply patient guidance of the Senior Publisher, Rob Langham. In addition, two anonymous reviewers provided valuable feedback on the proposal for this volume, which usefully expanded the scope of the project and impacted a number of the conversations among authors. Christopher Grimsley assisted in a variety of technical aspects of production.

Editing this collection has been a labor of love, and I am especially grateful to those who focused my passion for this subject, inspired me, and saw me through this project: Jill E. Millstone, without whom I might still be idling in the depths of quantum chemistry, blissfully unaware of just how complex and wondrous material behavior could be an order of magnitude higher; Bob Batterman, who is never as interested in classification as I am and who repeatedly cautioned me against editing a collection of essays, but who nonetheless encourages me to so many of my finish lines; Michaela Massimi, who first sparked my interest in philosophical theories of kinds; John Norton, who challenged me to articulate what made chemistry different from other sciences; Mark Wilson, Jim Woodward, and Nora Boyd, for helpful critiques of my own views; the National Science Foundation's Graduate Research Fellowship Program, for providing financial support for the research that originated this effort under grant #1247842; Michael Hartmann, Patrick Straney, Lauren Marbella, Chris Andolina, Ashley Smith, and the rest of the Millstone Lab for their interdisciplinary collaboration; the contributors to this volume, particularly Catherine Kendig, Melinda Fagan, and Vadim Keyser, for their encouragement and support; and Jennifer Jhun, Aleta Quinn, Elizabeth O'Neill, and Kathryn Tabb, for the accountability that is pressing these words to the page. Above all, I am grateful to Jeffrey Bursten Sykora for sustaining my strength throughout this process and in all things.

Historical lenses on classification in chemistry and biology

2 Crafting names and making kinds

Lessons from the 1892 Geneva Nomenclature Congress

Evan Hepler-Smith

Introduction

On the evening of Easter Monday, 18 April 1892, thirty-four chemists assembled in the Hôtel Métropole, at the southwestern tip of Lake Geneva where it empties into the Rhône, looking out over the Jet d'Eau and across the lake to the mountains to the north. The group included presidents of national chemical societies, editors of prestigious journals, and three future Nobel laureates, gathered for an event that a local newspaper promised would "mark an important date in the history of chemistry" (Chronique Locale. Congrès de Chimie, 17 April 1892). They had come to Geneva to address a serious obstacle to the continued progress of chemical science and industry: confusing nomenclature. Over the following four days – cold, but pleasant, the Congress secretary noted in his diary – in morning and afternoon sessions of about three hours each, the delegates of the Geneva Congress traversed the field of organic chemical substances and mapped it out in names (Pictet, 1892b).

What's in a nomenclature? In the case of organic chemistry, the answer is both complex and simple. The substance of organic chemical nomenclature is complex: thousands of pages of byzantine rules for generating millions of names, most of which sprawl out illegibly over dozens of syllables (see, e.g., Favre and Powell, 2014 (1568 pages); Chemical Abstracts Service, 1982 (approx. 2000 pages)). Underlying this complexity, however, is a simple principle: the systematic name of a compound should express its chemical structure (Figure 2.1).

More specifically, systematic chemical names are one-to-one mappings of structural formulas, diagrams prized by chemists as their "graphic language" for representing chemical substances as networks of atoms linked by bonds (Hoffmann and Laszlo, 2012). Systematic nomenclature is a compromise: chemists accept ungainly chemical names and obscure nomenclature rules as a "necessary evil" that makes it possible to identify and order chemical substances according to the much-loved structural formulas (Loening, 1985).

This relationship between name, diagram, and substance was established at the Geneva Congress.[1] The sixty-two rules codified there were to be the foundation of a general method for the "faithful translation" of structural formulas into names. The Geneva rules christened only a small fraction of then-known organic

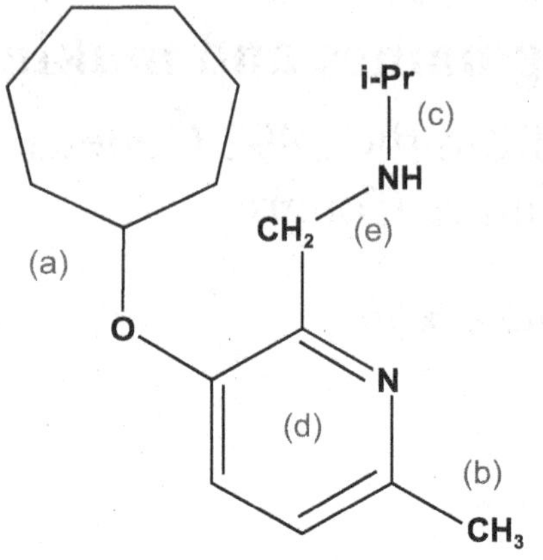

3-(Cycloheptyloxy)-6-methyl-*N*-
(1-methylethyl)-2-pyridinemethanamine

a) 3-(Cycloheptyloxy)   d) -2-pyridine
b) -6-methyl   e) methanamine
c) -*N*-(1-methylethyl)

Figure 2.1 The structural formula and systematic name of the seventy-five millionth small
molecule added to the Chemical Abstracts Service Registry

Source: "CAS REGISTRY Surpasses 75 Million Small Molecules," Press Release of 11 November 2013,
https://www.prnewswire.com/news-releases/cas-registrysm-surpasses-75-million-small-molecules-
231463401.html.

compounds with systematic names, and awkward names, at that; after an initial
burst of enthusiasm, most chemists saw them as excessively rigid and narrow in
scope. Nevertheless, contemporary methods of organic chemical nomenclature
have their roots in the principles that the Geneva delegates formulated and followed in establishing these rules.[2]

The significance of the Geneva Congress lay not in a set of chemical names,
but in a conception of chemical nomenclature. As the pace of organic synthesis
accelerated during the 1870s and 1880s, conflicting methods for naming organic
compounds emerged. New synthetic organic substances – some, such as the azo
compounds, prized as potential dyes or pharmaceuticals – accumulated multiple
different names, and these chemical synonyms became an impediment to chemists seeking to keep track of their rapidly developing field.[3] However, many of
these names shared a common basis: the arrangement of chemical subunits that
made up the compound, as understood according to the principles of structure
theory. Synonymy made nomenclature reform desirable, the chemical industry
made it valuable, and structure theory made it conceivable.

None of these, however, made it happen. That was the work of Charles Friedel.
While his German peers put structural formulas to use in their race to outdo each

other's synthetic achievements, the Alsatian savant battled opponents in the Parisian academic bureaucracy over the legitimacy of structure theory and its place in French classrooms. In this context, Friedel and his scientific allies reconceived nomenclature not simply as a collection of names but a field that spanned all organic substances and that might be made to express experimentally attested structural relationships between them. They saw an international effort to bring about such a system of nomenclature as an opportunity to advance French chemistry and their own position within it.

The reform that the Geneva Congress undertook, however, would be guided by a different vision of how, and where, chemical names might be put to work. German chemist Adolf von Baeyer convinced his fellow delegates that their nomenclature should comprise a set of rules for generating unique, "official" names tailored for use in indexing chemical journals and reference works. In carrying out this plan, the Geneva Congress invented the sort of name, and articulated the relationship between name, diagram, and substance, that has distinguished chemical nomenclature ever since.

The relationship between chemical diagrams and chemical substances has justly received the careful attention of historians and philosophers (see, e.g., Klein, 2003; Nye, 1993; Rocke, 2010; Slater, 2002; Dagognet, 1969). These scholars have illuminated the manipulability, visual suggestiveness, and semantic density that have made graphical representations such fruitful "paper tools" for organic chemists over the past two centuries. Chemical nomenclature since Lavoisier, in contrast, has received comparatively little recent scholarly attention, as Bernadette Bensaude-Vincent has noted (Bensaude-Vincent, 2003, p. 175).[4] For historians as for chemists, rules of organic chemical nomenclature can be technical, tedious, and seemingly far removed from the epistemic shifts, instrumental revolutions, and transformations of the material world that have characterized the modern chemical sciences. Yet the tedium and technicality of systematic names have been the price that chemists have paid to maintain the utility of their favourite diagrams and the cumulative development of their science. Neither of the latter can be fully understood apart from the former, and their relationship was forged at the Geneva Congress.[5]

"The language itself must be transformed"

The international collection of chemists who gathered at Geneva was well suited to taking action on matters of organic chemical nomenclature (Table 2.1). Nearly all were specialists in organic chemistry, and many, including their host, Carl Graebe, had already taken special interest in matters of naming. Baeyer and Friedel, leading figures of the field, had the professional stature to lend authority to the work of the Congress. So did rising stars like Emil Fischer, who took advantage of the convenient timing and location of the Congress to spend a week with his former teacher Baeyer, relaxing and "practicing a little French" in a Swiss resort town before continuing on to Geneva (Fischer, 1892).[6] Ferdinand Tiemann and Maurice Hanriot, editors of the influential journals of the German and French

Table 2.1 List of delegates to the Geneva Congress, after the roster printed in the Congress proceedings

Allemagne.	*MM.*
MM.	A. Combes (Paris).
A. von Baeyer (Munich).	C. Friedel (»).
E. Fischer (Würzburg).	A. Haller (Nancy).
E. von Meyer (Leipzig).	M. Hanriot (Paris).
E. Noelting (Mulhous).	A. Le Bel (»).
F. Tiemann (Berlin).	L. Maquenne (Paris).
Angleterre.	*Hollande.*
H.-E. Armstrong (Londres).	A.-P.-N. Franchimont (Leyde).
J.-H. Gladstone (»).	
W. Ramsay (»).	*Italie.*
Autriche.	S. Cannizzaro (Rome).
A. Lieben (Vienne).	A. Cossa (Turin).
Z. Skraup (Graz).	M. Fileti (»).
	E. Paterno (Palerme).
Belgique.	
M. Delacre (Gand).	*Roumanie.*
	C. Istrati (Bucarest).
France.	
A. Arnaud (Paris).	*Suisse.*
Ph. Barbier (Lyon).	C. Graebe (Genève).
A. Béhal (Paris).	P.-A. Guye (»).
L. Bouveault (Paris).	A. Hantzsch (Zurich).
P. Cazeneuve (Lyon).	D. Monnier (Genève).
	R. Nietzki (Bâle).
	A. Pictet (Genève).

Source: Pictet, 1892a, "Congrès International," p. 487.

chemical societies, attended; Friedrich Beilstein, compiler of the exhaustive, indispensable *Handbuch der Organischen Chemie*, was unable to make the journey from St Petersburg, but contributed his approbation and several suggestions. (Gordin, 2005; Pictet, 1892a). Such editors were in a position both to communicate the Congress's decisions on nomenclature to a broad audience and, more important, to enforce them in the pages of their publications.[7]

The official proceedings of the Congress – and historical accounts that follow it – describe a shared aim that united this august group in pursuit of nomenclature reform: the desire to bring order to a chemical lexicon set in disarray by the rapid accumulation of new compounds and alternative ways of naming them (Pictet, 1892a, pp. 485–486).[8] However, this common purpose was a product of the reform effort that culminated in the Geneva Congress, not its cause. During the 1880s, two sets of chemists each confronted a different problem of chemical language. Chemists studying complex synthetic compounds, predominantly in German settings, were struggling with the various synonyms used by different authors to identify these compounds. In contrast, a faction of Parisian chemists led by Friedel was striving to incorporate terminology and notation at the heart of their approach to chemistry into the highly regulated French chemical curriculum. Though nomenclature reform brought these two groups together, their goals and perspectives remained distinct.

Both the proliferation of chemical synonyms and Friedel's hopes for pedagogical reform were rooted in the principles of structure theory. Structure theory offered a means of interpreting and predicting chemical phenomena in terms of constitution: the particular patterns in which atoms of elements combined within a molecule. The classification of *in*organic compounds according to their composition had been a central feature of late eighteenth-century chemistry.[9] By the early nineteenth century, however, chemists had found that the composition of organic compounds often did not provide a reliable basis for establishing their identity or relationships. There were even chemical twins, termed *isomers*, which contained the same elements in the same proportions but displayed different chemical properties.[10]

Consequently, Lavoisier's composition-based binomial nomenclature for inorganic compounds could not be applied to organic chemical substances (Crosland, 1978, pp. 133–224). Names based in a theory of constitution could, but during the 1850s, there were many such theories, as the diverse chemical formulas used to express their conclusions illustrated.[11] Some forms of "rational nomenclature" came into use nonetheless, but more often, chemists retained and coined names for organic compounds based on their sources or properties (Crosland, 1978, pp. 285–318). A substance isolated from ants was "formic acid" (Latin: *formica*); an oil whose vapour caused an acid-soaked splint to burst into bright red flame, "pyrrol" (Greek: *pyrros*, red, fiery + Latin: *oleum*, oil, as in *benzol*); the product of the *dehyd*rogenation of *al*cohol, "aldehyde" (Flood, 1963, pp. 29–30, 95, 187).

From the late 1850s on, a group of imaginative chemists drew together certain insights from the various constitutional theories of the preceding decade, forming the productive and ever more broadly shared principles of structure theory.[12] These included *atomism*, the assumption that substances were made up of chemically indivisible atoms whose relative weights could be derived by means of certain physical measurements and laws; *substitution*, which divided each molecule into a core subunit belonging to its "parent compound" and one or more "substituents" in place of hydrogen atoms of that parent; *valence*, the concept that atoms of each element formed a characteristic number of bonds to other atoms; the *tetravalence of carbon*, the rule that each carbon atom formed four bonds; and *self-linking*, the ability of atoms of the same element – especially carbon – to bond to one another.

As with previous approaches to constitution, descriptions and classifications based in structure theory could be expressed by conventions of rational nomenclature. August Wilhelm Hofmann fused older applications of prefixes and suffixes into such a scheme of hydrocarbon nomenclature, one of a few naming conventions that gained wide, though selective and informal, adoption. However, a new kind of chemical formula proved more influential.

During the early 1860s, Scottish chemist Alexander Crum Brown developed a style of graphic notation that expressed the constitution of compounds in accordance with the principles of structure theory.[13] Crum Brown's graphical formulas – soon thereafter termed "structural formulas" – represented atoms with element symbols like C, H, and O, and chemical associations between them with lines. Whereas most existing formulas privileged one feature of a compound, structural formulas provided a highly legible basis for identifying the various constitutional units that characterized organic substances and their relationships. By the end of the 1860s, Crum Brown's formulas were in broad use, helping to abate the preceding confusion over notation (Figure 2.2). Most users of structural formulas insisted that the diagrams were not meant to represent the physical microstructure of compounds, but they sometimes thought about chemical phenomena as if the formulas did.[14] As a shared, visually suggestive means of representing compounds in the idiom of structure theory, structural formulas were a productive aid to reasoning and teaching.

Structure theory developed in large part through, and in the service of, experiments in which chemists explored constitutional questions raised by valuable naturally occurring substances, using techniques of synthesis. The chemists who carried out such investigations hoped eventually to discover artificial methods of producing such compounds as indigo, morphine, and quinine.[15] However, the proximate goal of these "synthetical experiments" was not typically to prepare a singular target molecule, but to solve particular constitutional puzzles or determine the transformations effected by specific reactions.[16] This involved a shotgun approach: submitting numerous compounds to parallel reactions, and examining

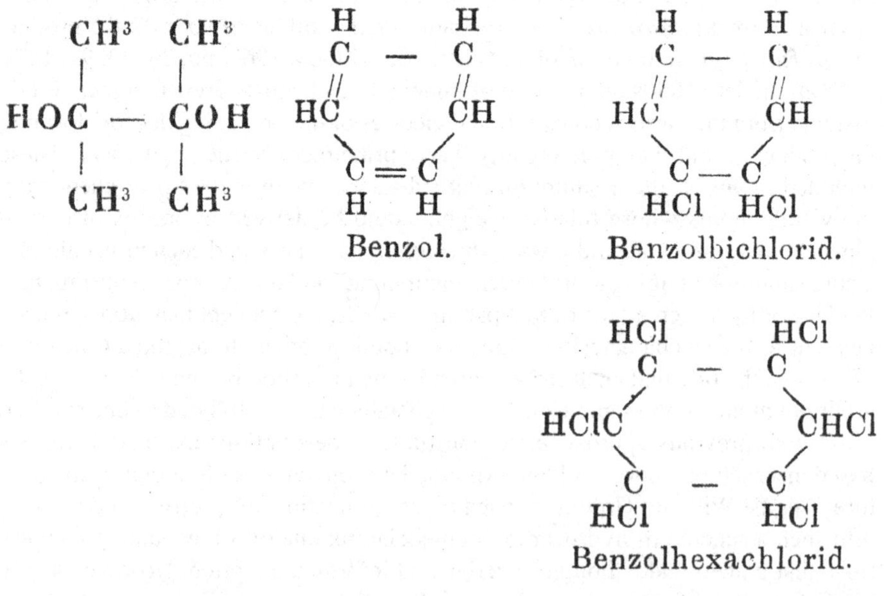

Figure 2.2 Structural formulas used by Geneva delegates Charles Friedel (left) and Carl Graebe (right) in articles of 1869 (Friedel, 1869, p. 397) and 1868 (Graebe, 1868, p. 68), respectively

the myriad synthetic products that resulted. Some of these novel substances turned out to be valuable dyes in their own right, and the synthetic dye industry that sprang up around these discoveries generated new constitutional puzzles for organic chemists to tackle (Travis, 1992). Such developments encouraged chemists to expand their synthetic studies, panning the by-products of their studies of reactions and constitution for chemically interesting or commercially valuable synthetic nuggets.[17]

These broad, comparative synthetic programs demanded considerable material resources, organization, and skilled chemical labour. From the 1860s through the 1880s, the conditions for such studies were by far most favourable in German universities, where professors of chemistry worked in well-equipped new laboratories flush with students ready to take up research projects.[18] With no central authority setting curricular standards, these professors could frame their research and instruction as they saw fit; most chose to predict, assess, and express the results of their synthetic studies using structural formulas.[19] The diagrams were at once an expedient means for making claims regarding the real constitution of a new synthetic product – chemists often simply combined the structural formulas of starting materials to illustrate the constitution of such compounds – and convenient bookkeeping devices for keeping track of the expanding taxonomy of organic chemistry.

The chemists who created these compounds sought to coin names that described and classified them in a similar manner. As one German chemist explained, such a procedure "eases comprehension in the highest degree, because the names are constructed entirely according to the formulas" (Heumann, 1882, pp. 813–814). Sometimes, existing conventions of rational nomenclature sufficed, as in the case of Fischer's derivatives of "triphenylmethane," a series of synthetic compounds sharing what he determined to be the core constitutional feature of the synthetic dye magenta (Fischer and Fischer, 1878). However, many synthetic experiments focussed on new reactions and newly determined structural units, producing compounds whose most salient features had no established name. In such cases, chemists often simply invented their own conventions for putting related fragments of structural formula into words.

Since different chemists often investigated the same classes of substances, this practice tended to produce synonyms. A method for naming one new compound could often be applied to numerous related substances; chemists found it particularly justifiable to do so when competing names had already accumulated, making it difficult to identify those compounds unambiguously. The spreading synonymy generated false structural analogies and obscured intended ones, undermining the ability of these names to convey the constitutional relationships they had been coined to express. The situation was most dire in the areas of greatest chemical interest. Surveying the terms that three dozen chemists had applied to a class of alkaloid constituents and their close chemical relatives, the German-trained Swede Oscar Widman found it "very difficult to keep track of all of these nearly innumerable names, chosen according to different principles, for compounds that are closely related to one another in their constitution, or are even identical" (Figure 2.3) (Widman, 1888, pp. 186–187).

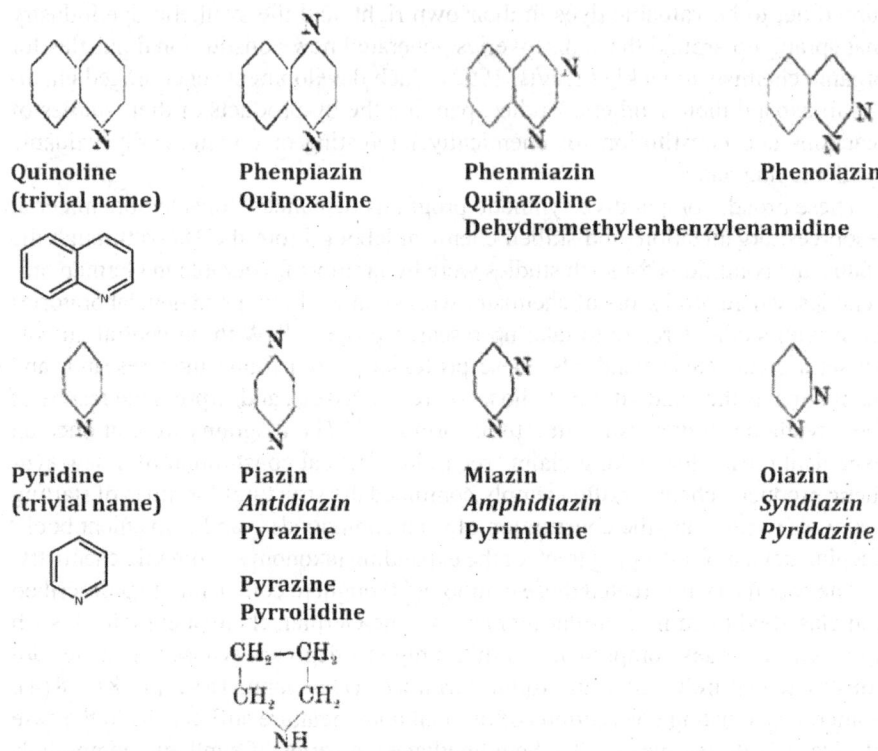

Figure 2.3 Competing rational names for some nitrogen-containing ring compounds

Note: All names and original diagrams are from Widman (1888) except for names in italics, from Hantzsch (1891). Structural formulas for pyridine and quinoline, drawn according to present-day conventions, are provided for reference. Unlabelled vertices of structural formulas indicate carbon atoms, and hydrogen atoms are omitted, except for the last.

Widman and others presented schemes to bring order to the nomenclature of chemical families mired in such confusion (Baeyer, 1884; Hantzsch, 1888). However, these individual proposals rarely mitigated the synonymy and sometimes added to it. No chemists working in the competitive, decentralized, resource-rich settings of German chemistry chose to leave their laboratories to launch sustained, collective action on the matter.

Charles Friedel faced a different challenge.[20] In 1884, thirty years after entering the laboratory of Adolphe Wurtz, Friedel succeeded his deceased mentor as Professor of Organic Chemistry at the Sorbonne. Wurtz and his students had been important early contributors to structure theory; over the succeeding decades, the best students formed a close-knit group of French savants who practiced and promoted this approach to chemistry, led by Wurtz and his protégé Friedel. In France, scholarship was centralized in Paris, education tightly regulated by a hierarchical academic bureaucracy, and positivism an influential source of criteria for judging scientific claims. Matters of notation and terminology were particularly hotly debated.[21] Promoting structure theory, therefore, meant engaging in polem--ics with sceptical rivals and taking care to establish a firm empirical basis for structural formulas.

To Wurtz and Friedel, who saw their native province of Alsace lost to the newly unified Germany in the Franco-Prussian war of 1870–1871, their intellectual cause was also a patriotic one. Many French savants and statesmen blamed the disastrous defeat on the decline of French science and technology relative to their German counterparts. Wurtz and Friedel contended that only through embracing atomism and structure theory (and upgrading the nation's poorly equipped academic laboratories) could French chemical science and industry advance.

By the 1880s, Wurtz, Friedel, and some of their students managed to introduce structure theory into their teaching, especially in the provinces, but their goal of establishing it within the general French chemical curriculum remained elusive. Friedel remarked:

> Although science advances fairly rapidly, public education, in its prudent march, struggles to follow this movement, especially in our country, bound to uniformity in its very organization. . . . [T]he language itself must be transformed to express the new ideas, and nothing is more difficult to accept than a change of language.
>
> (Friedel, 1886, pp. xxv–xxvi)

In 1888, Friedel seized the following year's Universal Exposition in Paris as an opportunity to begin such a transformation. He organized an International Congress of Chemistry, to be held during the Exposition; one of the sections of the Congress addressed the "unification of chemical nomenclature and notation" (Minutes of Council, 1888).[22] Chemists from nine nations participated, but the vast majority were French, and Friedel and his students dominated the nomenclature discussions.[23] They also took the reins of the International Commission formed to continue the study of nomenclature in preparation for a future Congress. This Commission, chaired by Friedel, included representatives from fourteen nations, but it immediately delegated the task of developing specific proposals to a specially designated Subcommission. The Subcommission consisted of Friedel; his fellow former Wurtz students Armand Gautier and Édouard Grimaux; three of Friedel's own students, Alphonse Combes, Auguste Béhal, and Adrien Fauconnier; and Louis Bouveault, a student of Wurtz disciple Maurice Hanriot (Combes, 1892, p. 258n1).

Two years and forty-five meetings later, the Subcommission prepared to submit its report to a different kind of Congress. For the Geneva Congress, in contrast to the Paris meeting, Friedel took pains to secure the participation of an influential collection of chemists, including some of the most prominent Germans struggling with the challenge of synonymy.[24] Such a group taking up such carefully crafted proposals could dare to hope that its collective effort to reform chemical nomenclature might succeed.

Yet these delegates brought distinct motives and perspectives to this collective effort. Friedel and his French colleagues wished to reform the language in which chemistry was taught, while Baeyer and his fellow Germans wanted to find a more certain means of documenting the identity of chemical substances and the relationships among them. Although the delegates had similar ideas regarding the epistemic reliability and limits of structure theory, the use of structural formulas was a matter of habit as well as epistemology. Encouraged by the rapid pace of synthetic chemistry, and enabled by their relative scholarly and pedagogical autonomy, German chemists used structural formulas confidently, both as tools of visual reasoning, as if they were mimetic representations, and as bookkeeping devices for cataloguing chemical subunits. The French atomists were no less attached to structural formulas, but in the disputatious arena of Parisian science, they had to be ready to fend off challenges to the legitimacy of these diagrams. In order to do so, they were accustomed to keeping structural formulas in close contact with the experimental evidence on which they were based.[25] At the Geneva Congress, these separate motives for reform and practices of graphical representation shaped two distinct visions of how and where systematic nomenclature should work.

Making nomenclature official

On Tuesday morning, the delegates climbed the hill of the old city to the Salle du Grand Conseil for the Congress's opening session (Réunions-Convocations-Concerts, 19 April 1892). Friedel began the morning's proceedings by introducing the fifty-three page report that his Subcommission had prepared. The report expressed a bold reinterpretation of the Commission's original mandate. Instead of seeking to unify the nomenclature of particular families of compounds, Friedel's group had conceived an entirely new understanding of nomenclature that would unite the names of all organic chemical substances under a system governed by a few key principles. Before the morning's session was through, however, Friedel's plan would be supplemented and in part supplanted by Adolf von Baeyer's vision of how and where such a system of names should work.

The Paris Congress had charged the International Commission it created with unifying the uncertain nomenclature of three groups of complex compounds.[26] In its report, the Subcommission explained that in order to carry out these tasks, it found itself compelled to consider two further matters:

d   The rules to follow in naming each chemical function by a consistent suffix or a prefix inserted into the name;
e   The designations to apply to radicals and in general to functional groups most often present in molecules.[27]

(Rapport de la Sous-Commission, 1892, pp. 393–395)

The Subcommission members decided that they could not unify the nomenclature of complex substances without first setting out rules for naming their simpler component parts. This meant, however, that the Subcommission's proposals would address not only the substances at the leading edge of synthesis that had prompted the reform effort, but also simple compounds with well-established names. Friedel and his colleagues were open about this aim: "the propositions adopted by the Subcommission, at the end of numerous meetings . . . allow, we believe, the clear, succinct, and unambiguous naming of all known organic chemical substances" (Rapport de la Sous-Commission, 1892, p. 393). Friedel and his Subcommission had been tasked with unifying the outer reaches of chemical vocabulary, but they proposed a universal grammar for forming acceptable chemical names.

The Subcommission framed seven principles to guide its pursuit of this ambitious goal:

1   Change current nomenclature as little as possible, except to complete and standardize it in order to avoid confusion and superfluous synonyms.
2   As much as possible, found the laws of the reformed nomenclature on the general principle of *substitutions*, except where it's necessary to express a direct addition.
3   In all of the names of derivatives of the same family – of the same hydrocarbon, the same alcohol, etc. – include a shared root that indicates their relationship and their parent.
4   Express the groups which characterize the functions of these molecules by prefixes and suffixes added to this root.
5   Conserve established names such as camphor, xylene, naphthalene, and terebine as roots, without indicating the structure of these familiar compounds or radicals.
6   As much as possible, adopt or construct names that are spoken and written according to the chemical formula, separating out each radical and indicating their positions following a fixed order.
7   Conserve common names such as alcohol, camphor, chloral, quinine, indigo, tyrosine, and urea, since they have passed into regular usage, but without prejudice against the corresponding new names.[28]

(Rapport de la Sous-Commission, 1892, pp. 393–394)

This was a blueprint for nomenclature reform that took into account both the benefits of rational nomenclature and the drawbacks of replacing established terms with neologisms. The second and sixth of these principles tethered the new nomenclature to structure theory and its diagrams. The name of an organic compound would follow the architecture of its structural formula, divided according to the principle of substitution into substituent radicals and a core corresponding to a parent compound. The third and fourth principles articulated the linguistic form in which the reformed names would express chemical relationships. Substitution products derived from the same parent compound would share a root, and compounds possessing the same chemical function would share a prefix or suffix. These were familiar conventions for forming rational names, but they had not previously been stated formally or applied generally.

Decades of experience had taught the French reformers how resistant their colleagues could be to changes in notation that appeared unjustified. Friedel and his band, therefore, took care to rein in their reformist impulses with the first, fifth, and seventh principles, which expressed a commitment to retaining established names that were not direct sources of confusion. Collectively, the Subcommission's principles described a reform that was sweeping in scope but accommodating toward habit, sanctioning chemical names that the Subcommission members hoped would prove "universally accepted, simple, clear, and lasting" (Rapport de la Sous-Commission, 1892, p. 398).

Previous proposals addressing organic chemical nomenclature had always addressed the naming of some particular set of compounds. So, too, did the questions that Friedel's Subcommission had been charged with answering. *Nomenclature* was always succeeded by *of*; nobody spoke of organic chemical nomenclature in general. Friedel and his Subcommission changed this. The scheme that Friedel presented at the outset of the Geneva Congress redefined nomenclature as a collective lexicon addressed at once to all organic substances. Order was to be established within this field in three ways: through a unit-by-unit correspondence between name and structural formula, through establishing such correspondence in a consistent fashion across many different classes of substances, and through a conservative commitment to preserving familiar names that had proven to be clear and useful.

Having laid out this approach to nomenclature reform, Friedel opened the floor for discussion of the particular recommendations that made up the bulk of the report. Baeyer, however, was not yet ready to move on. He requested that, prior to entering into a detailed discussion of any particular family of compounds, the Congress answer a few basic questions that had not yet been settled to his satisfaction. Which settings required nomenclature reform most urgently? What principles did the Congress need to prioritize in order to ensure that the new names were effective in these settings? How far was the reform to extend? (Pictet, 1892a, pp. 489–490, 1892c, p. 170).

Friedel and his fellow Subcommission members had sought to accommodate the various demands that might be placed on names by building flexibility into their scheme. They proposed allowing chemists to select among various nomenclature procedures that could apply to a compound (Rapport de la Sous-Commission, 1892, pp. 397–398). In response to Baeyer's challenge, Friedel and his colleagues remarked that this freedom of choice was especially useful in pedagogical settings, enabling an instructor to highlight the various chemical functions of a compound using different names. They cited the example of aniline, a compound of vital importance in chemical research and manufacturing. Different aspects of the chemical behaviour of aniline were associated with two structural units, an amine functional group and a benzene core. By referring to the substance as *phenylamine* or *aminobenzene*, an instructor could emphasize either of these properties and the chemical relationships that it entailed (Pictet, 1892a, p. 489).

Baeyer, however, was primarily concerned with establishing a reformed nomenclature for the purposes of chemical lexicography. In his view, the demands of indexing were so pressing that the Congress should focus *exclusively* on establishing for each compound "an official name, the translation of its constitution, which would permit it to be retrieved easily in tables and dictionaries" (Pictet,

1892c, p. 170).[29] The length and euphony of these official names would be relegated to secondary considerations. In journal articles, lectures, classrooms, and other settings where such concerns were significant, chemists would remain free to use whatever names they wished.

After a long discussion with Lieben, Bouveault, and Friedel, Baeyer won the assent of the President and the other delegates to his plan. The Congress agreed unanimously to adopt Baeyer's version of systematic nomenclature as its first resolution: "Alongside the usual procedures of nomenclature, an official name will be established for each organic compound, permitting it to be found under a unique heading in indexes and dictionaries"(Pictet, 1892a, pp. 490–491). With that, they broke for lunch.

The decision amounted to a division of the field of nomenclature which the Subcommission had defined as the object of its proposals. The notion of a field encompassing the names of all organic substances remained, cast as "the usual procedures of nomenclature." Following Baeyer's lead, the Congress carved out a separate field of "official names," whose application was restricted to reference works. The Subcommission's vision of order, however, no longer applied to either field. Within the compass of indexes, Friedel's commitments to lexical conservatism and flexibility were trumped by the imperative of forming unique names.[30] Elsewhere, on the other hand, the Congress would not attempt any reform at all.

The Congress secretary Amé Pictet summarized:

> We will continue to say alcohol, chloroform, sugar, and the chemist who, either in his articles or in the classroom, would use the expressions "ethanol," "trichloromethane," and "hexanal-pentol," which are henceforth the official names of these three substances, would fall into the same ridiculous pedantism as the botanist who in normal conversation employed the term "brassica oleracea" to designate "un chou" [a cabbage].
>
> (Pictet, 1892c, p. 168)

The structure of chemical function and the function of chemical structure

The group reconvened that afternoon at the Hôtel Métropole, where the remaining six sessions of the Congress were to be held. Having established their aim – a set of rules for constructing a unique, official name based on the constitution of each organic compound – they had to work out how to achieve it (Pictet, Agenda, entries of 19–21 April). This was a question of the relationship between name, formula diagram, and substance. Once again, the Geneva delegates arrived at a different answer than that proposed by the Subcommission. In doing so, the Congress articulated a distinctive way of thinking about structural formulas that shaped the use of these diagrams and helped bring about their remarkable persistence.

The first item on the session's agenda was the saturated hydrocarbons: compounds made up of carbon atoms linked in open chains, with all of their remaining units of valence taken up by hydrogen atoms. They were commonly named in several different ways, but these names were well-established and caused little confusion. In its report, the Subcommission paid little attention to these simplest and least reactive of organic compounds.

Tasked with forming official names, the Geneva delegates had to be more particular. They quickly agreed that Hofmann's well-known names – *ethane*, *pentane*, and the like – should be uniquely assigned to the "normal" isomers whose carbon atoms were arranged in a linear chain. As for the branched isomers, Lieben thought that the Subcommission's approach to a different set of compounds might be of use. For the special case of open-chain compounds containing multiple functional groups, the French chemists had adopted a more rule-bound alternative to their flexible method. Béhal, the author of this portion of the report, proposed taking the longest linear chain of carbon atoms in the compound's carbon "skeleton" as a starting point, numbering the carbon atoms of this chain from one end to the other. The name of the hydrocarbon corresponding to this chain would serve as the root of the name of the compound. Each functional group would be indicated by a prefix or suffix, along with the number of the skeleton carbon atom to which it was attached (Rapport de la Sous-Commission, 1892, pp. 414–416).

In Lieben's view, branched hydrocarbons could be named by an analogous process: locating the longest carbon chain in the structural formula, taking the name of the hydrocarbon corresponding to this "principal chain" as a root, and then naming the remaining "side chains" as if they were functional substituents (Pictet, 1892a, pp. 493–494). Baeyer suggested adopting Béhal's method of numbering for the branched hydrocarbons, as well. In view of the imperative of generating unique names, their fellow delegates agreed to the propositions.

Baeyer had further plans for these simple compounds. He called on the Congress to make the rules for naming hydrocarbons the basis of its entire system, applying them as a first step in naming any organic compound. Béhal disagreed, proposing instead to consider the position of functional groups as the starting point for determining a compound's name. He contended that this would make for a more straightforward naming process, shorter names, and – most important – names that more clearly expressed the compound's chemical behaviour. Baeyer countered that unlike such "functional nomenclature," his procedure would ensure consistent naming and numbering among different derivatives with the same carbon skeleton (Béhal, 1892).

After some debate, the delegates approved Baeyer's proposal. As a result, no matter what the substance, the Geneva rules latched on to it through the "ensemble of atoms of carbon directly bonded to one another, forming an invariable skeleton, which is found in all the compounds derived by substitution of the hydrocarbon that contains it" (Combes, 1892, p. 258). The Geneva rules generated consistent, unique names through the application of a consistent, even algorithmic procedure: take a compound's structural formula, reduce it to a carbon skeleton, identify the longest chain in that skeleton, and so on through fourteen rules of hydrocarbon nomenclature before moving on to functional groups. The Geneva Nomenclature was systematic in process as well as product.

For example, take pinacone, a compound produced by the coupling of two molecules of acetone. When Friedel studied this substance in the 1860s, in addition to its established trivial name, Friedel assigned it the name *tetramethyl ethylglycol*. He selected the root "glycol" in order to emphasize the compound's chemical function – a set of properties and characteristic chemical reactions that Friedel had established through painstaking experiment. The Subcommission's report advocated names that directly expressed such analogies in chemical behaviour (Friedel, 1869, pp. 322, 390–397).

In contrast, a chemical editor applying the Geneva rules to name pinacone would begin with its structural formula (Figure 2.4). Setting aside the two alcohol functional groups, he would identify the longest chain within the carbon skeleton and select the name of the saturated hydrocarbon corresponding to this four-carbon chain, *butane*, as a root (Figure 2.4, I). Treating the two remaining carbon atoms as side chains, he would use the single-carbon substituent prefix *methyl-*, adding *di-* to account for both of them, forming *dimethylbutane* (Figure 2.4, II). After

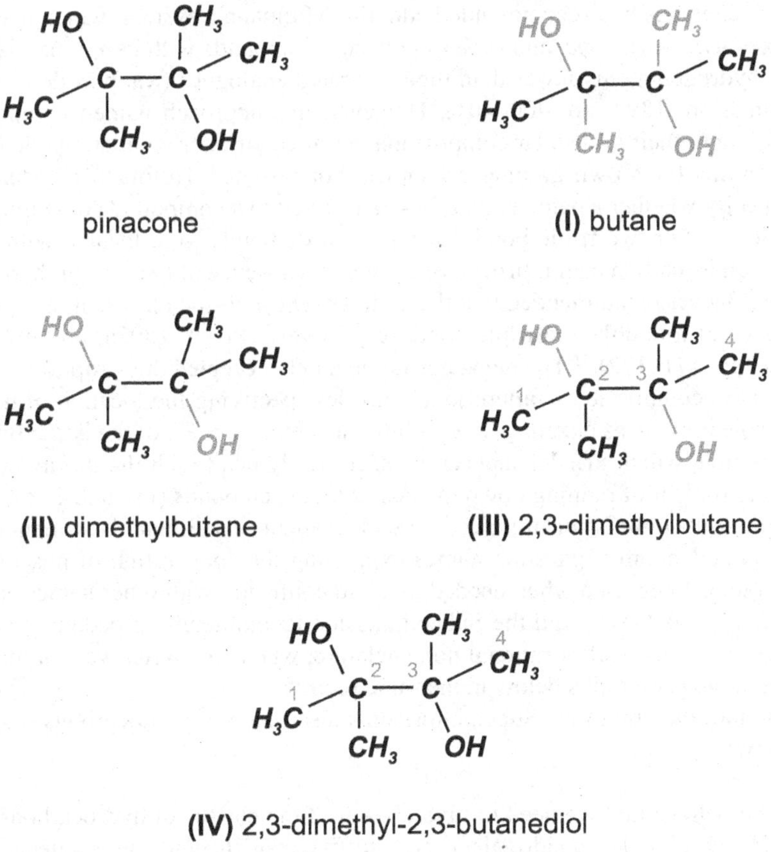

pinacone

(I) butane

(II) dimethylbutane

(III) 2,3-dimethylbutane

(IV) 2,3-dimethyl-2,3-butanediol

Figure 2.4 Naming pinacone according to the Geneva rules

numbering the principal chain, he would arrive at the full name of the hydrocarbon skeleton, *2,3-dimethylbutane* (Figure 2.4, III). With the skeleton taken care of, he would then turn to the -OH groups. Adding *di-*, for the two groups, to the functional suffix *-ol*, and using the numbering determined by the hydrocarbon skeleton, he would complete the official name for pinacone, *2,3-dimethyl-2,3-butanediol* (Figure 2.4, IV). At no point in the process was empirical study of the properties of the compound to be taken into account. Instead, the relationship between the Geneva name of an organic compound and its functional behaviour and relationships was mediated by the Geneva rules and the structural formulas on which they operated.

The difference between the Subcommission's and the Congress's views on the relationship of formula, name, and substance was illustrated more sharply still in the treatment of unsaturated hydrocarbons, in which one or more pairs of adjacent carbon atoms were linked by two or three bonds rather than "saturated" with bonds to hydrogen atoms. As Pictet reported it in the proceedings of the Congress, Baeyer's proposal for naming unsaturated hydrocarbons was an incontestable correction of a surprisingly fundamental oversight by the Subcommission. The French chemists had recommended adopting Hofmann's system, which used the suffixes *-ene, -ine, -one*, and *-une* to indicate compounds with two, four, six, and eight hydrogen atoms fewer than their saturated analogues (Rapport de la Sous-Commission, 1892, pp. 400–401). However, this approach named compounds according to their molecular composition, not their structure, violating one of the Subcommission's own guiding principles. For instance, Hofmann's system did not specify whether a name ending in *-ine* referred to a compound containing one double bond or two triple bonds, or where those bonds were located within the compound's carbon chain. Instead of the general absence of two or four hydrogen atoms, Baeyer recommended that the suffixes *-ene* and *-ine* represent the specific presence of a double and triple bond, respectively (Pictet, 1892a, pp. 497–498, 1892c, pp. 171–172).[31] The delegates immediately accepted this proposal.

The Subcommission's failure to adopt rules specifying the location of double and triple bonds was puzzling. It wasn't for lack of awareness of the issue: Béhal's dissertation, which Friedel supervised, specifically dealt with the thorny experimental problem of pinning down the position of such bonds (Friedel, 1889). Further, the Subcommission followed its endorsement of Hofmann's nomenclature with an addendum suggesting names expressing the constitution of unsaturated compounds to be used when needed to avoid confusion with other names ending in *-ine* or *-one*.[32] Why did the Subcommission recommend a procedure that did not live up to its goal of rational nomenclature, when an alternative that did was sitting three paragraphs below in the same report?

The introduction to the Subcommission's discussion of hydrocarbons suggests an answer:

> We believe that we need to give to each of the classes of hydrocarbons, saturated, bivalent, quadrivalent, etc., suffixes which mark their valence, this having been measured by the number of atoms of chlorine or bromine or

double the number of molecules of hydracids that each hydrocarbon is able to accumulate in its molecule, *without substitution and without the molecule losing its ability to return to its original state.*

(Rapport de la Sous-Commission, 1892, p. 400. Emphasis in original.)

For the Subcommission, the valence of an unsaturated hydrocarbon was grounded not in a feature of a structural formula, but in experiment. The procedure described earlier offered a straightforward method for determining valence, whereas it could be difficult to determine the specific location of a double- or triple-bond in a carbon chain, as Béhal and Friedel knew first-hand. If a compound's constitution could be established in the former manner but not the latter, the Subcommission provided a mechanism for assigning it a name that expressed as much of its constitution as had been determined. Without a unique, complete structural formula, however, such a compound could not be named at all by the Geneva rules.[33]

A further reason for this approach was epistemic caution. Friedel and his collaborators wished to avoid creating names that expressed a degree of structural detail not warranted by experiment. They wrote:

Instead of employing interminable, obscure names, often founded on dubious hypotheses and incomplete and uncertain proofs, names having the pretension of expressing definitively the entire constitution of the compound, it would be better, in general, to limit ourselves to expressing clearly by the name of the substance its relationships with the other substances of the natural family to which it is known to belong and its characteristic and experimental functional properties.

(Rapport de la Sous-Commission, 1892, p. 398)

The Subcommission members saw the value in names that expressed "the entire constitution" of compounds. But heeding the tenuousness of many determinations of chemical structure, they took care not to reach beyond experimentally attested properties and relationships in order to do so.

"What is important above all," Pictet wrote in summarizing the decisions of the Geneva Congress, "is that the official name be the faithful translation of the constitution of the compound, and that it represent the constitution to the mind just as the structural formula does" (Pictet, 1892a, p. 491). Over their two years of labour preparing for the Congress, Friedel and his fellow Subcommission members had also sought to develop chemical names that worked "just as the structural formula does." The distinct relationships that the Subcommission and the Congress hoped to forge between name, diagram, and substance reflected the differing goals of each group.

Friedel and his allies aimed to reinforce the teaching of structure theory in France in a manner that could resist the dogged criticism of atomic sceptics. To that end, they crafted proposals for a nomenclature that expressed experimentally determined chemical behaviour through the interpretive lens of structure theory, just as their structural formulas did. According to the Subcommission's proposal, systematic names and formula diagrams would each represent organic compounds in the same manner, grounded ultimately in observations of chemical function.

Baeyer's goal, in contrast, was to bring the advantages of structural formulas to bear on a problem to which the diagrams themselves couldn't be applied: determining a unique heading for each compound in an alphabetically ordered index. Guided by Baeyer and Lieben, the Geneva delegates built a set of rules for systematically disassembling a structural formula and naming and numbering its pieces in a determined order, such that the process could be reversed to regenerate the diagram from the official name. The proximate referent of a name formed according to the Geneva rules was not the compound at all, but the structural formula. Geneva names stood in for structural formulas where the latter could not be used and summoned these diagrams, when needed, to represent the compounds themselves.

A "solid and durable foundation"

The discussion of the open-chain hydrocarbons occupied the entire Tuesday afternoon session and most of Wednesday morning; the resulting set of rules constituted the general framework of the Geneva Nomenclature. The approach departed from the spirit of the Subcommission's report, but Friedel and his cohort did not express dissatisfaction – at least not on the record. After all, if they had conceded the battle for the form of the new nomenclature, they were winning the war of convincing the assembled chemists to agree upon *some* collection of nomenclature rules.

This achievement was possible in large part through the intimacy and collegiality of the Geneva Congress. British delegate Henry Armstrong wrote of the Congress's opening days:

> The great advantage to be derived from the personal intercourse which such meetings promote was soon apparent: gradually, the doubts which many entertained as to the possibility of devising a practical rational scheme of nomenclature were dispersed, and ere many hours had elapsed the sympathies of all present were enlisted on behalf of the work.
>
> (Armstrong, 1892, p. 57)

Much of this "personal intercourse" took place outside of the Congress's twice-daily official sessions. After dining together on Tuesday evening, the chemists proceeded to the theatre, where the Geneva minister of education invited them to join him in his lounge and box. The evening's show, the well-known fairy comedy *Le Pied du Mouton*, allowed the delegates to focus their attention on conversation and champagne.[34]

The next evening, the delegates assembled for the official Congress banquet, joined by several of their wives and a number of local dignitaries.[35] As the guests dined on a main course of "sirloin à la nomenclature," Graebe toasted Friedel for his leadership in the nomenclature reform effort. Friedel, in turn, toasted Graebe, Geneva, and Switzerland for their amiable welcome (Chronique Locale. Congrès de Chimie, 22 April 1892). Baeyer toasted the successive roles played by France and now Germany in driving the discipline of chemistry forward, and drank to whichever nation would next claim the laurels (Armstrong, 1892, pp. 56–57). The evening concluded with a nightcap at a nearby haunt, the *Taverne du Crocodile* (Pictet, 1892b).

Reflecting on the meeting years later, Emil Fischer contrasted the series of "noisy and confusing" congresses of applied chemistry inaugurated in 1894, which he "avoided if at all possible," with the Geneva Congress, whose distinctive social and intellectual dimensions made the thirty-four chemists in attendance feel "like one big family" (Fischer, 1922, p. 135). Such ties also encouraged the delegates to put their professional authority behind the resulting Geneva rules, uniting the assembly as "a mission . . . which will explain the enterprise to chemists generally," as Armstrong put it (Armstrong, 1892, p. 57).

In order to take advantage of these favourable circumstances, however, the Congress had to work quickly. As Wednesday drew to a close, the delegates had dispatched only a fraction of the simple organic compounds containing a single functional group, and Baeyer, Lieben, Fischer, and others were set to depart the following day. Stanislao Cannizzaro, one of the Congress vice-presidents, proposed nominating another commission to deal with the many matters that seemed sure to remain unresolved (Pictet, 1892c, p. 175). But leaving Geneva without at least a reasonably complete system of nomenclature would mean missing the opportunity to establish the new names in the forthcoming edition of Beilstein's *Handbuch* and other settings in which they could be disseminated and tested.

When Friedel brought up the acid anhydrides as the final order of business on Wednesday, Graebe proposed that the "current mode" of designating these substances be retained (Pictet, 1892c, p. 175). This set the pattern for the Congress's final two days, in which the delegates switched gears from intensive debate to churning out resolution after resolution, either rubber-stamping Subcommission recommendations or conserving "the current nomenclature." A dispute between Friedel and Baeyer over what actually constituted one such "current practice" might have given them pause.[36] However, any such qualms seem to have been outweighed by the imperative of bringing as many compounds as possible within the compass of the Geneva rules. In their final substantive meeting on Friday morning, after half of the delegates had departed, the remaining attendees approved Combes's long proposal for numbering benzene derivatives in its entirety, without any discussion (Pictet, 1892b). They were persuaded, Pictet wrote, by the "simplicity and precision of the notation that he proposed." Exhaustion, the lack of time for further debate, and a determination to broaden the officially sanctioned nomenclature as far as possible probably played just as great a part (Pictet, 1892a, pp. 512–519).[37]

For several chemical families, including compounds with more than one variety of functional group, the delegates were unable to reach a consensus. Nor could they elect to sanction current practice, for these compounds had no well-established names. Further, the delegates decided that they could not responsibly ignore the demands of general usage when any rule they ratified would be the only available standard for such usage. In reserving the naming of such compounds for further study, the Congress called for an "attempt to reconcile the exigencies of spoken nomenclature with those of a terminology applicable to dictionaries" (Pictet, 1892a, pp. 509–511, 519–520). Choosing to create a nomenclature expressly for dictionaries had freed the delegates from worrying too much about cumbersome names that its rules might create; they were just for indexes, after all, not for use in other settings. For this reason, however, the Geneva delegates came up short where generally recognized names had not yet been established – precisely those sets of compounds whose nomenclature the International Commission had originally been tasked with unifying.

This irony did not much bother the Congress attendees. They had come to Geneva not to criticize nomenclature reform, but to carry it out. They carefully limited the scope of their project to "official" nomenclature and prioritized the development of consistent rules to determine a unique name for any compound whose structure was relatively simple. They set aside for further study questions of how to extend these rules to determine unique names for more complex compounds, including most of those that Friedel's Commission had been created to address. According to Pictet, the assembled chemists felt that, despite the limited scope of their achievements,

> the results achieved so far are nonetheless of considerable importance, and the International Congress of Geneva, in establishing the principles of the official nomenclature, will have laid a solid and durable foundation for the reform that it intended to accomplish.
>
> (Pictet, 1892a, p. 520)

The delegates departed Geneva satisfied that their rules of nomenclature were complete enough to be applied and robust enough to build upon.

Conclusion

The Geneva Congress did more than introduce a new set of names for organic chemical compounds, or even a new way of naming them. Over four days in April 1892, Europe's greatest authorities on organic chemistry crafted a new conception of chemical nomenclature, and a new relationship between chemical names, structural formulas, and chemical substances.

Shifts in how chemists used and coined names during the 1870s and 1880s had made nomenclature reform conceivable. The practice of basing the names of substances on their structural formulas provided a means; the subsequent proliferation of confusing synonyms provided a motive. But the Geneva Congress actually came to pass through the considerable efforts of Charles Friedel, who was distinctively positioned to see nomenclature reform as an opportunity to advance the causes of France in chemical science and industry and of structure theory in France.

In laying the groundwork for Geneva, Friedel and his band of students and associates redefined nomenclature not simply as the collected vocabulary of chemists but as a continuous field that could be made subject to a flexible grammar grounded in the constitution of chemical compounds, just as structural formulas were. Guided by lexicographic rather than pedagogical concerns, Adolf von Baeyer convinced the delegates of the Geneva Congress to redefine nomenclature once again, dividing it into a sphere of general usage to be left to its own devices and a realm of official nomenclature, where each name was a precise and unique transcription of a structural formula diagram.

Neither Friedel nor Baeyer intended this form of nomenclature to replace well-established trivial names, and indeed, it never did.[38] But the correspondence of name to structural formula, a proximate relation into which the substance itself did not enter, quickly became the characteristic feature associated with chemical names.[39] Alexander Crum Brown, who had been among the first chemists to use structural formulas, recognized this. Discussing the Geneva Nomenclature two months after the Congress, he observed, "We must keep in mind that such systematic names as had been suggested were really names of formulas rather than names of substances" (Chemical Society Meeting of 16 June, 1892).

This entailed not only a different kind of chemical name but a different manner of reading a structural formula. The rules of the Geneva Nomenclature did not operate on the chemical properties and relationships that the formula expressed, but on regularities inherent to the diagrams: patterns of connection among atomic symbols that provided purchase for a logical order of operations mapping diagram to name. This way of thinking about structural formulas would subsequently shape not only such "paper tools" as Robert Robinson's electron displacement arrows, but also print, mechanical, and electronic technologies for managing chemical information.[40]

That was all to come. The immediate response to the Geneva Nomenclature was a flurry of adoption, followed by a rising tide of scepticism. Pictet published his French-language account of the Congress and its resolutions in May 1892. Translations of the Geneva rules in English, German, Italian, Russian, and Romanian soon followed (Verkade, 1985, pp. 8, 51).[41] Beilstein adopted Geneva names in his *Handbuch*, to the delight of the French (Roussanova, 2007 vol. 2 pp. 405–407). Notwithstanding the Congress's decision to restrict the application of its rules to reference works, some delegates proposed using or adapting them for teaching (Béhal, 1892, p. 411; Armstrong, 1892, pp. 57–58). One even wrote a textbook that used exclusively Geneva names (Istrati, 1893).[42]

Meanwhile, the chemical editors for whom the Congress had tailored its resolutions were growing frustrated with the Geneva Nomenclature. The rules were complicated, the names long and awkward, and most of the compounds of interest to their authors and readers were not even covered. Some, such as the new editors of the *Berichte*, the influential journal of the German Chemical Society, chose to stop using the official names at all.

The German editors based their decision on the relationship that the Geneva approach forged between name, diagram, and substance – and on the effects of *this* relationship on relations among chemists:

> The rational index-nomenclature must, in order to avoid arbitrariness, treat all compounds as equals, as the citizens of a socialist State would be treated. The [non-index] text-nomenclature – corresponding to our existing social

order – raises out from the great masses certain compounds to which fate or merit has assigned a "role."

(Jacobson and Stelzner, 1898, p. 3372)

With the accumulation of more and more rules, and more and more official names indistinguishable to those without special training in nomenclature, these editors worried that fewer and fewer chemists would be capable of using the chemical literature effectively: "Such a state of affairs – a separation of chemists into index-educated scholars and index-ignorant technicians – would be highly undesirable and would contradict the traditions of our science" (Jacobson and Stelzner, 1898, p. 3375). The radical *egalité* that the Geneva Nomenclature imposed upon the metaphorical polity of organic substances would, feared the editors of the *Berichte*, bring about a real division within the social order of chemistry.

They were not entirely wrong. On the one hand, by enshrining the late-nineteenth-century structural formula as the basis of chemical names, the Geneva Congress helped establish the status of these diagrams as the "iconic vernacular" of chemistry (Hoffmann and Laszlo, 2012, p. 176). On the other, the Congress also launched the study of methods for "official" nomenclature as a separate channel of research running parallel to the chemical mainstream. As the size of the chemical world and the complexity of compounds increased, systematic nomenclature grew ever more esoteric and ever more vital. Within international commissions, chemical publications enterprises, manufacturers' laboratories, and elsewhere, chemists continued to chase the elusive goal set at Geneva: generating systematic names – or machine-readable ciphers, or digital tables – that represent compounds just as the structural formula does.

Acknowledgements

This chapter is based on research supported by the Science History Institute and the Society for the History of Alchemy and Chemistry. This chapter was originally published as "'Just as the Structural Formula Does': Names, Diagrams, and the Structure of Organic Chemistry at the 1892 Geneva Nomenclature Congress," by Evan Hepler-Smith, *Ambix.* Vol 62:1 pp. 1–28 (2015), copyright © Society for the History of Alchemy and Chemistry, reprinted by permission of Taylor & Francis Ltd, www.tandfonline.com on behalf of the Society for the History of Alchemy and Chemistry.

Notes

1 I refer to structural formulas as "diagrams" in order to highlight their *visual* mode of representing facts or claims about chemical constitution, which the chemists who are the subject of this essay sought to express using words. I make no claim here regarding the relationship between other sorts of visual representation and "these little diagrams," as Hoffmann and Laszlo (2012, p. 164) call them.
2 P. E. Verkade, longtime chairman of the IUPAC commission on organic nomenclature, presents a detailed technical analysis of both the limitations and influence of the Geneva

Nomenclature in his account of the Congress, by far the most thorough historical treatment of its intellectual subject matter (Verkade, 1985, pp. 1–48, 276–298). Other works that discuss the Geneva Congress in surveying the development of chemical nomenclature during the nineteenth and twentieth centuries include Crosland (1978, pp. 347–354), Bensaude-Vincent (2003), Traynham (1987, 1993), and Grignard (1935).

3 I use "synonym" here and throughout this essay in its taxonomic sense, that is, to refer to one of multiple different terms that specifically refer to the same object. This is the sense in which the participants in the Geneva Congress used the term, for example, in a discussion of how best to avoid the problem of the "multiplicity of synonyms" (Rapport de la Sous-Commission, 1892).

4 For exceptions, see the works mentioned in note 2, especially (Crosland, 1978) and (Verkade, 1985), as well as (Dagognet, 1969).

5 A note on translation: for the sake of intelligibility, I have translated French and German chemical names, affixes, roots, and orthography into the corresponding forms used in contemporaneous English. The Geneva Congress was conducted primarily in French; the official text of its rules was published in French, as well. The Congress left any decisions regarding how to implement its rules in other languages to the individual authors who applied or translated them. Though the delegates to the Geneva Congress did not address the implications of linguistic difference for chemical nomenclature, others did, especially in discussions that addressed spoken chemical names as well as written ones. I address such matters and provide a more detailed account of the emergence and subsequent history of systematic organic chemical nomenclature in (Hepler-Smith, 2016).

6 The British delegate Henry Armstrong noted that some delegates had difficulty expressing or understanding the intricacies of nomenclature questions in French (Armstrong, 1892).

7 There had been no such mechanism for the enforcement of the decisions of the Karlsruhe Congress, an 1860 gathering in which 140 chemists – including Baeyer, Friedel, and three other Geneva delegates – had gathered to discuss chemical terminology and notation. Though memorable and indirectly influential, the Congress had little direct effect on chemists' practice. See (Carlsruhe Compte Rendu, 1984; Rocke, 2001, pp. 226–233; Bensaude-Vincent, 1990, 2003).

8 This "lexicon" took material form in organic chemical reference works, for example, (Richter, 1900).

9 On the "theory domain of composition" in eighteenth-century chemistry, see (Kim, 2003, pp. 65–110). On "compositionism," see (Chang, 2012, pp. 1–70, defined on 14).

10 On the emergence of an experimental culture of organic chemistry in the first half of the nineteenth century, see (Klein, 2003, pp. 41–85).

11 In a textbook fascicle originally published in 1859, August Kekulé dramatized the variety in theories of chemical constitution proposed over the preceding decade with a table of nineteen different formulas for acetic acid (vinegar) (Kekulé, 1861, p. 58).

12 Alan J. Rocke has chronicled this history in illuminating detail. On priority claims regarding structure theory, see (Rocke, 1981). On the numerous chemists whose work made structure theory conceivable, see (Rocke, 1993). On the role of imagination in attempts to plumb the chemical micro-world, see (Rocke, 2010).

13 Crum Brown probably drew inspiration from the diagrams used by his countryman A. S. Couper in 1858. Crum Brown's were the most influential among several styles of graphical formula developed during the early 1860s to express constitution in terms of structure theory. Their success was probably due both to their visual suggestiveness and their comparability to the type formulas in use for the previous decade (Rocke, 2010, pp. 118–160).

14 Whatever their particular commitments with regard to epistemology and chemical theory, as a matter of practice, the majority of nineteenth-century chemists took on this sort of position, which Rocke has termed "chemical atomism" (Rocke, 1984).

15 In 1868, Carl Graebe and Carl Liebermann achieved the first commercially significant natural product synthesis. Their preparation of alizarin, the primary colourant in madder dye, was a pivotal episode in the development of the synthetic dye industry. (Travis, 1993, pp. 163–190).

16 As Catherine Jackson has shown, Hofmann introduced "synthetical experiments" of this sort during the early 1840s and continued to rely on this mode of research for the rest of his career. Jackson argues that chemists who first succeeded in "constructive synthesis" – the deliberate construction of complex target molecules – during the 1880s did so by means of skilful and innovative laboratory practice, not because of special theoretical insights or particular uses of formula diagrams (Jackson, 2014a, 2014b).

17 Beginning in the 1870s, dye firms began to establish their own research laboratories where they specifically pursued the discovery of such compounds (Travis, 1993, pp. 209–230). Physical chemists invoked the prospecting metaphor in critiquing this mode of organic chemical research (Servos, 1990, p. 64).

18 See, for example, (Johnson, 1985; Borscheid, 1976, pp. 16–82; Rocke, 1993, pp. 270–286, 2001, pp. 392–397). Both before and after German unification, the chemical institutes of many Austrian and Swiss universities partook of the same academic culture, if not always such generous state sponsorship.

19 The broad adoption of structural formulas among German organic chemists during the 1870s is well illustrated by the vigorous but isolated campaign of Hermann Kolbe against their use (Rocke, 1993, pp. 325–339).

20 The following two paragraphs draw upon (Rocke, 2001), (Fauque, 2003), and (Carneiro and Pigeard, 1997).

21 See, for instance, the detailed discussion of debates over the meaning of the terms "atomicity," "equivalent," "atom," and "molecule," and of controversies over notation, in (Fauque, 2003).

22 On international scientific congresses in the late nineteenth century, see the essays collected in (Schroeder-Gudehus, 1990).

23 Untimely distribution of invitations reportedly contributed to the modest international attendance (The International Chemical Congress, 1889). Whether accidental or intended, the paltry international showing helped Friedel and his students maintain control of the proceedings.

24 For example, eight of the best-known chemists mentioned in Widman's article were invited; three attended.

25 On the speculative use of structural formulas by German organic chemists, see (Rocke, 1993, pp. 325–339).

26 Specifically, compounds with rings containing nitrogen or other non-carbon atoms (Widman's area of concern), compounds with multiple different functional groups, and benzene derivatives.

27 A "functional group" was (and is) understood as a discrete structural subunit within a chemical formula that was associated with a particular chemical behaviour.

28 Those interested in fine gradations of language will note that the Subcommission distinguished technical names that refer to a chemical compound without specifying its structure, mentioned in the fifth of these principles, from chemical names either derived from or passed into regular nontechnical usage, mentioned in the seventh. This distinction corresponds approximately to the distinction between *trivial* and *common* chemical names, though the former class arguably includes both sets of names.

29 Prior to the Geneva Congress, Baeyer submitted a proposal for determining unique names to the Subcommission; the Subcommission considered but rejected this proposal, along with an alternative prepared by Subcommission member Louis Bouveault. However, the Subcommission members had not understood Baeyer to be proposing that the reformers *restrict* their efforts to establishing such a system of official nomenclature (Rapport de la Sous-Commission, 1892, p. 394).

30 Pictet accordingly edited the Subcommission's list of guiding principles for his official report, eliminating those that mentioned conserving existing names. Compare (Rapport de la Sous-Commission, 1892, pp. 393–394) and (Pictet, 1892a, p. 491).

31 Since the two suffixes were indistinguishable when pronounced in English, the suffix "ine" was later exchanged for "yne."

32 Specifically, those of alkaloids such as morphine, quinine, and nicotine, and ketones such as acetone.

33 The Congress explicitly restricted the scope of its rules to compounds whose structure had been determined (Pictet, 1892a, p. 492).

34 As Pictet put it in his diary, "Le pied de mouton, pièce à grand spect mais pas spir. . . . Passé presq tt mon temps avec eux dans le salon du cons d'Etat à boire du Champ. Longue conv avec Em Fischer." (Pictet, 1892b) *Le Pied du Mouton* (literally, "The Sheep's Foot") was a French comedy written in 1806, featuring slapstick comedy, fairies, and spectacular effects. It was popular throughout the century, especially with middle- and lower-class audiences, and was adapted as a film in 1907 (Zipes, 2011, pp. 37–38).

35 At the turn of the twentieth century, the wives of scientists often accompanied their husbands on professional travel. On the road, as at home, both spouses participated in constituting a gendered lifestyle that shaped the production of scientific knowledge (Bergwik, 2014).

36 The disagreement pertained to the usual meaning of the prefixes "sulfo" and "thio" (Béhal, 1892, p. 414).

37 The delegates did elect to set aside some difficult questions regarding the position of benzene in the Geneva rules' order of operations.

38 Present-day international nomenclature guidelines sanction the use of certain trivial names, called "retained names" (Favre and Powell, 2014, p. 9).

39 "To be useful for *communication* among chemists, nomenclature for chemical compounds should additionally contain within itself an explicit or implied relationship to the structure of the compound, in order that the reader or listener can deduce the structure (and thus the identity) from the name" (emphasis in original) (Panico et al., 1993, p. xiii).

40 On Robinson's arrows: (Nye, 1993, pp. 191–192). On early approaches to identifying compounds in mechanical and electronic environments, see (National Research Council Committee on Modern Methods of Handling Chemical Information, 1964).

41 The English translation, not cited by Verkade, is (Resolutions Adopted by the International Commission, 1892).

42 Friedel arranged for the publication of a French edition and wrote a preface for the translation.

References

Armstrong, H. (1892) The International Conference on Chemical Nomenclature. *Nature.* 46, 56–59.

Baeyer, A. V. (1884) Zur Chemischen Nomenclatur. *Berichte der Deutschen Chemischen Gesellschaft.* 17, 960–963.

Béhal, A. (1892) La nomenclature chimique au congrès international de Genève. *Moniteur Scientifique.* 39, 407–417.

Bensaude-Vincent, B. (1990) Karlsruhe, Septembre 1860: l'Atome en Congrès. *Relations Internationales.* 62, 149–169.

———. (2003) Languages in Chemistry. In M. J. Nye (Ed.), *The Cambridge History of Science, Volume 5: The Modern Physical and Mathematical Sciences* (pp. 174–190). Cambridge: Cambidge University Press.

Bergwik, S. (2014) An Assemblage of Science and Home: The Gendered Lifestyle of Svante Arrhenius and Early Twentieth-Century Physical Chemistry. *Isis.* 105, 265–291.

Borscheid, P. (1976) *Naturwissenschaft, Statt und Industrie in Baden (1848–1914).* Stuttgart: Klett.

Carlsruhe Compte Rendu. (1984) In M. J. Nye (Ed.), *The Question of the Atom: From the Karlsruhe Congress to the First Solvay Conference, 1860–1911* (pp. 5–28). Los Angeles: Tomash.

Carneiro, A. & Pigeard, N. (1997) Chimistes Alsaciens à Paris au 19ème Siècle: Un Réseau, Une École? *Annals of Science.* 54, 533–546.

Chang, H. (2012) *Is Water H2O? Evidence, Realism, and Pluralism.* Dordrecht: Spring Verlag.

Chemical Abstracts Service. (1982) *Chemical Substance Name Selection Manual.* 2 vols. Washington, DC: Chemical Abstracts Service.

Chemical Society Meeting of 16 June. (1892) *Proceedings of the Chemical Society.* 8(114), 127–140.

Chronique Locale. Congrès de Chimie. (1892, April 17) *Journal de Genève*, p. 2.

———. (1892, April 22) *Journal de Genève*, p. 3.

Combes, A. (1892) Le Congrès International de Nomenclature Chimique. *Revue Générale des Sciences Pures et Appliquées.* 3, 257–260.

Crosland, M. P. (1978) *Historical Studies in the Language of Chemistry.* New York: Dover.

Dagognet, F. (1969) *Tableaux et Langages de la Chimie.* Paris: Éditions du Seuil.

Fauque, D. (2003) La réception de la théorie atomique en France sous le Second Empire et au début del la IIIe République. *Archives internationales d'histoire des sciences.* 53, 64–112.

Favre, H. & Powell, W. (Eds.). (2014) *Nomenclature of Organic Chemistry: IUPAC Recommendations and Preferred Names 2013.* Cambridge: Royal Society of Chemistry.

Fischer, E. (1892) Letter to Adolf von Baeyer, 22 March 1892, Box 36. *Emil Fischer Papers, BANC MSS 71/95.* The Bancroft Library, University of California, Berkeley.

———. (1922) *Aus Meinem Leben.* Berlin: Springer.

Fischer, E. & Fischer, O. (1878) Ueber Triphenylmethan und Rosanilin. *Justus Liebigs Annalen der Chemie.* 194, 242–303.

Flood, W. (1963) *The Origins of Chemical Names.* London: Oldbourne.

Friedel, C. (1869) Recherches sur les acétones et sur les aldéhydes. *Annales de la chimie et la physique.* 16, 310–407.

———. (1886) Introduction. In C. A. Wurtz, *La théorie atomique.* 4th edition. Paris: F. Alcan.

———. (1889) *Rapports sur les thèses de Doctorat d'État, 1881–1889.* Pierrefitte-sur-Seine: 800–802, AJ/16/5534 Archives Nationales.

Gordin, M. D. (2005) Beilstein Unbound: The Pedagogical Unraveling of a Man and His Handbuch. In D. Kaiser (Ed.), *Pedagogy and the Practice of Science: Historical and Contemporary Perspectives* (pp. 11–39). Cambridge, MA: MIT Press.

Graebe, C. (1868) Ueber die s. g. Additionsproducte der aromatischen Verbindungen. *Annalen der Chemie.* 146, 66–73.

Grignard, V. (1935) *Traité de chimie organique.* Vol. 1. Paris: Masson.

Hantzsch, A. (1888) Untersuchungen über Azole. Allgemeine Bemerkungen über Azole. *Annalen der Chemie.* 249, 1–6.

———. (1891) Zur Nomenclatur stereoisomerer Stickstoffverbindungen und stickstoffhaltiger Ringe. *Berichte der Deutschen Chemischen Gesellschaft.* 24, 3479–3488.

Hepler-Smith, E. (2016) *Nominally Rational: Systematic Nomenclature and the Structures of Organic Chemistry, 1889–1940.* (Doctoral dissertation) Princeton University.

Heumann, K. (1882) Die Nomenclatur complicirter Azoverbindungen. *Berichte der Deutschen Chemischen Gesellschaft.* 15, 813–816.

Hoffmann, R. & Laszlo, P. (2012) Representation in Chemistry (1991). In J. Kovac & M. Weisberg (Eds.), *Roald Hoffmann on the Philosophy, Art, and Science of Chemistry* (pp. 163–192). New York: Oxford University Press.

The International Chemical Congress. (1889) *Nature.* 40, 369–371.

Istrati, C. (1893) *Curs Elementar de Chimie.* Bucharest: C. Göbl.

Jackson, C. M. (2014a) The Curious Case of Coniine: Constructive Synthesis and Aromatic Structure Theory. In U. Klein & C. Reinhardt (Eds.), *Objects of Chemical Inquiry* (pp. 61–102). Sagamore Beach, MA: Sagamore History Publications.

———. (2014b) Synthetical Expeiments and Alkaloid Analogues: Liebig, Hofmann and the Origins of Organic Synthesis. *Historical Studies in the Natural Sciences.* 44, 319–363.

Jacobson, P. & Stelzner, R. (1898) Zur Frage der Benennung und Registrirung der organischen Verbindungen. *Berichte der Deutschen Chemischen Gesellschaft.* 31, 3368–3388.

Johnson, J. A. (1985) Academic Chemistry in Imperial Germany. *Isis.* 76, 500–524.

Kekulé, A. (1861) *Lehrbuch der Organischen Chemie.* Erlangen: F. Enke.

Kim, M. G. (2003) *Affinity, That Elusive Dream: A Genealogy of the Chemical Revolution.* Cambridge, MA: MIT Press.

Klein, U. (2003) *Experiments, Models, Paper Tools: Cultures of Organic Chemistry in the Nineteenth Century.* Stanford: Stanford University Press.

Loening, K. L. (1985) Foreword. In P. E. Verkade & S. Davies, *A History of the Nomenclature of Organic Chemistry* (pp. ix–x). Boston: Reidel.

Minutes of Council (1888) 31 July 1888, Procès-verbeaux de la Société Chimique de Paris, Cahier 2, 77–79. Archives de la Société Chimique de France.

National Research Council Committee on Modern Methods of Handling Chemical Information. (1964) *Survey of Chemical Notation Systems: A Report.* Washington, DC: National Academy of Sciences.

Nye, M. J. (1993) *From Chemical Philosophy to Theoretical Chemistry: Dynamics of Matter and Dynamics of Disciplines, 1800–1950.* Berkeley: University of California Press.

Panico, R., Powell, W. H. & Richer, J.-C. (1993) *A Guide to IUPAC Nomenclature of Organic Compounds: Recommendations 1993.* Oxford: Blackwell Scientific Publications.

Pictet, A. (1892a) Le Congrès International de Genève pour la Réforme de la Nomenclature Chimique. *Archive des Sciences Physiques et Naturelles.* 27, 485–520.

———. (1892b) Agenda, Entries of 19–21 April. Geneva, Z 306: Musée d'Histoire des Sciences.

———. (1892c) Minutes of the Geneva Congress. Bibliotheque de Genève, MS Fr. 3423.

Rapport de la Sous-Commission. (1892) *Association Française pour l'Avancement des Sciences: Compte Rendu.* 21, 392–455.

Resolutions Adopted by the International Commission. (1892) *Chemical News.* 65, 277–280.

Réunions-Convocations-Concerts. (1892, April 19) *Journal de Genève*, p. 3.

Richter, M. M. (1900) *Lexikon der Kohlenstoff-Verbindungen.* 2nd edition, 2 vols. Hamburg and Leipzig: L. Voss.

Rocke, A. J. (1981) Kekulé, Butlerov, and the Historiography of the Theory of Chemical Structure. *British Journal for the History of Science.* 14, 27–57.

———. (1984) *Chemical Atomism in the Nineteenth Century: From Dalton to Cannizzaro.* Columbus: The Ohio State University Press.

———. (1993) *The Quiet Revolution: Hermann Kolbe and the Science of Organic Chemistry.* Berkeley: University of California Press.

————. (2001) *Nationalizing Science: Adolph Wurtz and the Battle for French Chemistry.* Cambridge, MA: MIT Press.

————. (2010) *Image and Reality: Kekulé, Kopp, and the Scientific Imagination.* Chicago: University of Chicago Press.

Roussanova, E. (Ed.). (2007) *Friedrich Konrad Beilstein, Chemiker zwier Nationen.* Vol. 2. Hamburg: Books on Demand.

Schroeder-Gudehus, B. (Ed.). (1990) Les Congès Scientifiques Internationaux. *Relations Internationales.* 62, 111–211.

Servos, J. W. (1990) *Physical Chemistry from Ostwald to Pauling: The Making of a Science in America.* Princeton: Princeton University Press.

Slater, L. B. (2002) Instruments and Rules: R. B. Woodward and the Tools of Twentieth-Century Organic Chemistry. *Studies in the History and Philosophy of Science Part A.* 33, 1–33.

Travis, A. (1992) Science's Powerful Companion: A.W. Hofmann's Investigation of Aniline Red and Its Derivatives. *British Journal for the History of Science.* 25, 27–44.

Travis, A. (1993) *The Rainbow Makers: The Origins of the Synthetic Dyestuffs Industry in Western Europe.* Bethlehem; London: Lehigh University Press; Associated University Presses.

Traynham, J. G. (1987) The Familiar and the Systematic: A Century of Contention in Organic Chemical Nomenclature. In J. G. Traynham (Ed.), *Essays on the History of Organic Chemistry* (pp. 114–126). Baton Rouge: Louisiana State University Press.

————. (1993) Organic Nomenclature: The Geneva Conference 1892 and the Following Fifty Years. In M. Volkan Kisakürek (Ed.), *Organic Chemistry: Its Language and Its State of the Art* (pp. 1–8). New York: VCH.

Verkade, P. E. (1985) *A History of the Nomenclature of Organic Chemistry.* (trans. S. G. Davies). Boston: Reidel.

Widman, O. (1888) Zur Nomenclatur der Verbindungen, welche Stickstoffkerne enthalten. *Journal fur praktische Chemie.* 38, 185–201.

Zipes, J. (2011) *The Enchanted Screen: The Unknown History of Fairy-Tale Films.* New York: Routledge.

3 Biological kinds at the turn of the 20th century

Characters, genes, and species as theoretical elements

Aleta Quinn

> Mendelism is therefore just such a conceptual notation as is used in algebra or in chemistry. No one objects to expressing a circle as $x^2 + y^2 = r^2$. No one objects to saying that $BaCl_2 + H2SO_4 = BaSO_4 + 2HCl$. No one should object to saying that $DR + RR = 1DR + 1RR$. We push things into the germ cells as we place the dollars in the magician's hat. Hocuspocus! They disappear! Presto! Out they come again! If we have marked our money we may find that that which appears from the magician's false-bottomed hat is not the same as that which we put in. But it looks the same and is good coin of the realm. We have a good right therefore to poke our characters into the germ cell and to pull them out again if by so doing we can develop – not a true conception of the mechanism of heredity – but a scheme that aids in describing an inheritance.
>
> (East, 1912, pp. 633–634)

Introduction

Philosophers of biology have most usually discussed "natural kinds" in a sense that was articulated by Putnam and Quine, though its roots go back as least as far as John Locke. The view is that to be a member of a natural kind means to have a collection of core, essential properties common to all members of the natural kind. The core properties explain all the many observable commonalities shared by members of the kind. The classic example of a natural kind, at least as old as Locke's account, is a chemical element. This account of natural kinds has recently been treated as a kind of default against which alternatives can be compared – for example, Griffiths' (1999) attempt to "square the circle" by defining biological species as natural kinds with historical essences.

In this essay, I present some 19th and early 20th century alternatives to the view of natural kinds and chemical elements as fixed by essential properties. I briefly describe John Stuart Mill's account of natural kinds, and efforts by his commenters to reconcile that account with Mill's claims about causation. These attempts foreshadow Griffiths' arguments. I then turn to the use of the chemical element analogy by 19th century geneticists, focusing on William Bateson in particular. The relationship of analogy between chemical elements and biological species is more complicated than is typically presented, especially in 19th century conversations about kindhood. I argue that William Bateson's chemistry analogies are best understood as analogies to theoretical entities in the history and practice of chemistry. Bateson did not intend that the units of heredity answer to material units that behave in ways analogous to material atoms. Bateson's point was that biologists of his day should postulate a theoretical entity, basic to the science as elements once were to chemistry. These elements were reinterpreted during Bateson's lifetime and replaced by electrons, neutrons, and protons as basic units. Rather than comparing the properties of chemical units to those of unit factors, Bateson intended to highlight the role of a unit entity in scientific theorizing and methodology. Scientific conversations about classification in the 19th to early 20th centuries relied on a notion of a unit of classification, which Bateson understood to have specific characteristic constraints, and which contemporary philosophical accounts elucidated.

I begin with a short précis of Mill's account of natural kinds. Mill's work, which is foundational to philosophy discussions, is basically contemporaneous with the biological conversations in the rest of the chapter. Rather than shared possession of essential properties, Mill's account focuses on "gappiness" in nature, and as we will see, his contemporary philosophers attempted to reconcile this approach with his views on causation. Contemporary debates about the nature of biological kinds reflected this philosophical debate in scientific practice.

Natural kinds

Mill on natural kinds

Natural kinds are often thought to be identified via definite properties that are relevant to causal transactions. Causal laws are supposed to hold irrespective of spatiotemporal location: they describe regularities that hold throughout all time and space with respect to the specified entities. This generalizability implies that the identified subjects of causal laws are, in a sense, "immutable". While members of kinds may change, the defining properties by virtue of which entities participate in causal laws must remain the same. If gold is a natural kind, any individual sample of gold must bear the properties by virtue of which gold participates in the causal transactions described by causal laws. Any entity that bears these properties, regardless of its spatiotemporal location, must be gold. On this picture, we can conclude that natural kinds must have some core properties, possession of which is necessary and sufficient for an entity to belong to the natural kind. It is sometimes possible to identify the core properties on which kind membership depends and which are responsible for observable resemblances among kind members. Gold is a paradigmatic example: all individual pieces of gold share resemblances that are caused by shared possession of the atomic number 79. Observable resemblances include ductility, color, density, and behavior when subjected to specific chemical reactions.

John Wilkins (2012) ascribed the aforementioned set of views about natural kinds to J. S. Mill and argued that Mill's account of natural kinds collapsed the historical distinction between logical species and biological species. However, Mill clearly identified a concept of natural kinds that are not united by joint possession of core causal properties. Some of Mill's 19th century commentators were well aware of this point, which has largely been missed by recent commentators (but see Magnus, 2014, 2015). Carveth Read (1877) suggested that we can reduce all natural kinds to the effects of causation, but his suggestion was intended to revise Mill's view. Mill expressly denied that causal necessity suffices to bound natural kinds. Rather, he held that natural kinds are united by uniformities of coexistence that are not dependent upon causation (Mill, 1843b, p. 120).

Mill introduced the natural kind concept through a historical account of the origin of group names. Mill claimed that usually when we form a general name, we do so with particular attributes in mind, "because we need a word by means of which to predicate the attributes which it connotes" (Mill, 1843a, p. 160). But in some cases, we recognize a need to group entities without explicit reference to any particular resemblances. Mill conceived of natural kinds as groups of objects that resemble each other in very many ways; in fact, the objects share innumerably many attributes in common. Precisely because they share so many attributes, we recognize quite easily that some objects belong together and group them without paying attention to any specific subset of attributes.

On Mill's view, this is not a provisional epistemic gap about the identity of core properties. In the case of true natural kinds, there is no core group of causally basic properties. Mill held that the shared attributes of a natural kind cannot be reduced to some few attributes that cause all the other resemblances. Indeed, the irreducibility is diagnostic of true Millian kinds. "White things" do not form a natural kind, even though they share innumerable attributes, in a sense: the attribute that Englishmen call them "white", that Frenchmen call them "blanc", that they reflect more sunlight than do black things, and so on. But all of these properties can be reduced to effects derived from the property of whiteness itself. In contrast, over decades and centuries, biologists have discovered ever more carefully studied properties shared by the natural group "crows", and these properties cannot be reduced (logically or causally) to each other. Mill evidently held that the existence of natural kinds, whose members share innumerably many resemblances in common, is simply a brute fact about the universe. Mill expressed confidence that in particular some high-level divisions of biological taxa represent natural kinds.

In Mill's view, though we can form many propositions expressing uniformities among members of the kind, we do so via simple enumeration (Mill, 1843b, p. 134). We cannot form inductive laws about resemblances shared within kinds. The reason is that the uniformities linking members of a Millian natural kind are not causal. Millian kinds cannot be investigated via "a system of rigorous and scientific induction" because:

> The basis of such a system is wanting: there is no general axiom, standing in the same relation to the uniformities of coexistence as the law of causation does to those of succession. The Methods of Induction applicable to the ascertainment of causes and effects, are grounded upon the principle that

everything which has a beginning must have some cause or other . . . But in an inquiry whether some kind (as crow) universally possesses a certain property (as blackness), there is no room for any assumption analogous to this.

(Mill, 1843b, p. 126)

In Mill's view, science includes a process of ever-generalizing induction that produces causal explanation, but also investigation of kinds as a separate matter. As scientific investigation proceeds, some proposed kinds will be discovered to be groups whose shared features can be explained via causal property dependence. Yet Mill seemed confident that some kinds will be ultimate and united by shared possession of non-reducible properties.

Mill's commentators

What about the structure of the universe underwrites Mill's appeals to "uniformities of co-existence" that are separate from the law of causation? Carveth Read, philosopher of mind and of logic at Cambridge and then the University College London, analyzed and extended Mill's work in a series of treatises (Read, 1878, 1898; and subsequent editions). Read (1877) pressed Mill's distinction between uniformities due to causation versus uniformities among natural kinds. As Read explicated him, Mill held that scientists hypothesize both causal laws and classes of co-existing resemblances: "For every Law of Causation is the Definition of a Class of Causal Instances; and every Definition of a Natural Kind is a Law of Co-existence" (Read, 1877, p. 344). Read described induction as a "test" of the constancy of the relations so predicated:

> "The Induction of relations of Succession is governed by the Law of Causation; the Induction of relations of Co-existence is aided (much less effectively) by the doctrine of Natural Kinds."
>
> [C]omplete generalisation requires that one should be reduced to the other; and, as it is, we cannot hesitate to endeavour to reduce Co-existence to the effect of Causation.
>
> (Read, 1877, pp. 345–346)

According to Read, the doctrine of Natural Kind is "very inferior" to the Law of Causation as an instrument of investigation, and therefore we should work to reduce uniformities of co-existence to regularities governed by causation. Absent any explicit principle to justify the Laws of Co-existence, natural kinds seem to be residual uniformities that resist scientific explanation. Each kind represents an invariable grouping of properties apparently as a matter of brute fact that we can no more explain than the initial configuration of matter in the universe (Mill, 1843b, p. 44).

Fabian Franklin and Christine Ladd-Franklin, scholars of mathematics and mathematics and logic, respectively, at Johns Hopkins University,[1] suggested a way to causally ground Millian natural kinds. The distinction between natural kind and non-natural kind can be made on the basis of the type of causal connection that unites the group. Mill was on the right track, Franklin and Franklin (1888) claimed, in denying that natural kinds are united by a core group of properties on which all other shared properties depend directly. Instead, kinds share very many properties due to "a certain community of origin" (Franklin and Franklin, 1888, p. 84). The members of a natural kind share some common historical past, and the very many resemblances among kind members are due to this past causal story.

The expected common origin story is rarely recognized when natural kinds are first identified, and the historical story may remain unknown. Indeed, Franklin and Franklin claimed that their negative prescription "natural kinds share properties that are not causally reducible" suffices to identify kinds even if the reader disagrees with their positive claim, that a common origin unites kind members. Franklin and Franklin offered chemical elements as an example:

> [T]he fact that all portions of matter which possess a few of the properties of sodium do actually possess all the other properties of sodium forces upon us the conviction that either the qualities or the objects have a real connexion with each other. If the former is the case, the properties of sodium are deductions from its molecular constitution; if the latter, then sodium is in a very valid sense a Natural Kind . . . something very different from an arbitrary and "merely intellectual" class: and this, whether one agrees or does not agree with the present writers in regarding the connexion between the objects to reside in a certain community of origin.
>
> (Franklin and Franklin, 1888, p. 85)

Franklin and Franklin were unsure about whether all sodium objects in fact share a common origin. In1888, the picture seemed clearer in the case of zoological species:

> [I]n the case of the animals forming a species, it would be preposterous to suppose that all the common qualities might be explained deductively from a few of them. These, then, form a Natural Kind, in the sense in which we have used the term; and, in this case, community of origin has been sufficiently shown to be the true ground of the classification.
>
> (Franklin and Franklin, 1888, p. 85)

Many 19th and early 20th century philosophers and historians (e.g., Locy, 1915; Pouchet, 1853) considered biological species a reasonably well-understood theoretical entity (Amundson, 2005, p. 35). As will be seen in the next section, early 20th century practicing biologists were not so agreed on the nature and reality of species. Nineteenth century philosophers considered biological species sufficiently well established as to serve as a model for understanding chemical elements (e.g., Whewell, 1828, 1847). In the 20th century, the comparison would switch (Bather, 1927): the chemical elements are sufficiently well understood that they might serve as philosophical models for understanding biological species. Here, for example, is Bateson:

> If we may once more introduce a physical analogy, the distinctions with which the systematic naturalist is concerned in the study of living things are as multifarious as those by which chemists were confronted in the early days of their science. Diversities due to mechanical mixtures, to allotropy, to differences of temperature and pressure, or to degree of hydration, had all to be severally distinguished before the essential diversity due to variety of chemical constitution stood out clearly, and I surmise that not till a stricter analysis of the diversities of animals and plants has been made on a comprehensive scale, shall we be in a position to declare with any confidence whether there is or is not a natural and physiological distinction between species and variety.
>
> (Bateson, 1913a, pp. 15–16)

Biological species

Analogizing species to elements

In 1913, William Bateson claimed in his *Problems of Genetics* that it had become fashionable for biologists to deny that species exist at all. In Bateson's telling, the fact that taxonomists continued to use the species concept was thought to be an artifact of taxonomists' role as cataloguers – an almost embarrassing failure to bring cataloguing up to date with the scientific discovery that species are not real:

> But the problem how to name the forms and where to draw lines, how much should be included under one name and where a new name was required, all this was felt, rather as a cataloguer's difficulty than as a physiological problem. And so we still hear on the one hand of the confusion caused by excessive "splitting" and subdivisions, and on the other of the uncritical "lumpers" who associate together under one name forms which another collector or observer would like to see distinguished.[2]
>
> In spite of Darwin's hopes, the acceptance of his views has led to no real improvement – scarcely indeed to any change at all in either the practice or aims of systematists.
>
> (Bateson, 1913a, p. 10)

But Bateson insisted that the continued use of "species" was not simply an artifact (for contemporary uses of "artifact" in biological settings, see Chapter 4). He argued that the idea of species points at something real in nature, since species are visibly separated by discontinuous gaps. We must investigate these gaps rather than (in Bateson's allegation) cleverly arranging specimen drawers to try to hide the phenomena:

Almost always the collections are arranged in such a way that the phenomena of variation are masked. Forms intermediate between two species are, if possible, sorted into separate boxes under a third specific name. If a species is liable to be constantly associated with a mutational form, the mutants are picked out, regardless of the circumstances of their origin, from the samples among which they were captured, and put apart under a special name.

(Bateson, 1913a, p. 11)

According to Bateson, the species concept had not outlived its usefulness. Species were "fixed" in the same way that chemical elements were fixed; in order to make scientific progress, biologists must face up to the fixity of species, and research its physiological underpinnings. Recognition of basic, "fixed" unit elements was crucial to the progress of chemistry as a science. Just such an advance was made in biology when Linnaeus "fixed" biological species by declaring that they vary only within definable limits (Locy, 1915, p. 129; Sanson, 1900, pp. 24–29; Saint-Hilaire, 1859, p. 373). We must assume that species are (almost always) un-blending and immutable.[3] Later we might discover that these assumptions are not true in all circumstances. But the very process of achieving further scientific advance requires an assumption of theoretical unit elements.

To Linnaeus we owe the scientific acceptance of the supposedly literal interpretation of Creation, under which each species of plant and animal descends from one or a few originally created organisms. Earlier natural philosophers and theologians accepted transmutation and spontaneous generation, and interpreted God's act of Creation as conferring to the earth powers of production by means of which organisms spring forth.[4] Bateson briefly gave some examples of pre-Linnaean transmutation, but his historical account focuses on how the doctrine of the fixity of species came about, and what positive advances it enabled. Without the fixity of species, Linnaeus could not have outlined his system of classification, and without a concept and sketch of a "Natural System", Darwin could not have formulated his theory.

In support of these claims, I will briefly discuss some pre-Synthesis accounts of the history of the doctrine of the fixity of species. This discussion further supports Amundson's (2005), Wilkins' (2009), and Winsor's (2003, 2006) arguments that pre-Linnaean biologists accepted a variety of transformist views about species. Whereas Amundson (2005) provided primary references indicating pre-Linnaean belief in transmutation, Bateson did not need to provide such evidence. The fact that the fixity of species was accepted in the 18th century was familiar to Bateson's contemporaries.[5]

The history of history of biology

William Locy's *Biology and Its Makers* straightforwardly attributed the doctrine of the fixity of species, including the supposedly literal interpretation of the Creation story in Genesis, to Linnaeus: "Linnaeus first pronounced on the doctrine of fixity of species" (Locy, 1915, p. 129).

Locy reported that Linnaeus subsequently softened his views such that genera were the Created types. "Nevertheless, it was owing to his influence, more than to that of any other writer of the period, that the dogma of fixity of species was established" (Locy, 1915, p. 129).

Locy (1915, p. 418) claimed that the Jesuit Francisco Suárez (1548–1617) introduced the dogma of special Creation to theology.

Sanson (1900, pp. 24–29) attributed the doctrine of the fixity of species – which he states would come to be the doctrine of the Creationists – to Linnaeus, who developed it directly from Genesis. Woodruff summarized Linnaeus' contribution to biology:

> Linnaeus crystallized two dogmas – constancy and continuity of species – which permeated biology and reached, in slightly different form, their high water mark, indeed a reductio ad absurdum, in Agassiz's Essay on Classification a century later – as fate would have it, just a year before Darwin's Origin of Species.[6]

> (Woodruff, 1921, p. 263)

Henry Fairfield Osborn (1913) included both scientists and theologians in his history of biology. He reported that spontaneous generation and transmutation were accepted in one or another form by Augustine, Erigena, Roscellinus, William of Occam, Albert the Great, Aquinas, and Bruno. Osborn situated Suárez as the first theological special Creationist, with Suárez rejecting Augustine's claims that the earth received powers of producing species. However, Osborn (1913, p. 84) then noted that Suárez speculated that some species (e.g., the leopard) arose post-Creation through the crossing of original species.

There was some dispute on Suárez's views. Mivart (1871) identified Suárez as a transmutationist in order to argue that revered Church fathers might accept transmutation without committing heresy. Huxley's (1871) devastating "Mr. Darwin's Critics" argued that Suárez was a dyed-in-the-wool special Creationist, intolerant of any form of transmutation. Huxley dated acceptance of the fixist, six-day Creation account to Suárez, allowing that the earlier Church fathers were transmutationists. Huxley probably wanted to associate species fixism with Suárez, whose name in anti-Catholic circles carried negative associations (Huxley compared him to Judas Iscariot and Robespierre).[7]

This was not the first attempt to claim pre-Linnaean fixity of species for political reasons. Blainville (1845) pushed the fixity of species back to Albert the Great (ca. 1200–1280), and praised Albertus Magnus for this doctrine since they claimed that scientific knowledge would be impossible without it. Pouchet (1853) claimed that Albert the Great had not only fixed species, thereby rendering scientific knowledge possible, but moreover had invented the scientific conception of the scale of nature ("la série animale" – Pouchet, 1853, p. 278).[8]

Isidore Geoffroy Saint-Hilaire (1859) suggested that de Blainville, Maupied, and Pouchet tried in vain to claim greater antiquity for the doctrine of the fixity of species. His implication was that, as contemporary fixists, they sought a longer and perhaps better (by connection to a Church Father) pedigree for their "great doc-trine". After critiquing their case for Albert directly, Saint-Hilaire (1859, p. 373; my translation) declared: "The doctrine of fixity dates scientifically, whatever has been said on the matter, neither from the middle ages, nor from the 19th century; neither from Albert the Great, nor from Cuvier; but from Linnaeus".[9]

Of the aforementioned authors, only Huxley and Saint-Hilaire were critical of the establishment of the fixity of species. The others lauded Linnaeus for his development of the species concept, counting fixity as an advance. But early 20th century writers blamed Linnaeus' followers for taking the concept too far and leading to the 19th century "hardening" of the species concept, to which Bateson referred: "The fixity of species was taken for granted, but not till the overt proclamation of evolutionary doctrine by Lamarck do we find the strenu-ous and passionate assertions of immutability characteristic of the first half of the 19th century" (Bateson, 1913a, p. 9). Bateson urged biologists to accept and investigate the fixity of species, yet he clearly did not advocate a return to the hardened, absolutist fixism of 19th century Creationists. The concept of fixed species – groups of organisms separated by discrete gaps from other groups – was crucial for scientific investigation. But this fixity should not be taken to deny that species evolve.

In this vein, the analogy to chemistry is especially helpful. Before elements were fixed, alchemists investigated a hopeless profusion of classifications and transmu-tations. Lavoisier developed the concept, of "element" to mean a type of substance that does not transform into any other type of substance. To identify a sample of substance as *sodium chloride* was to posit that there is sodium and chlorine in the sample, and that no matter what one did to the sample, sodium and chlorine would remain. Burn, dissolve, melt, or swallow the sample, the sodium and chlorine have to go somewhere, and the sodium chloride may be recoverable by subsequent processes.

Only after the elements were fixed was there progress in delineating the regular-ities of chemical reactions and combinations, progress that led to what early 20th century biologists perceived as a natural system: the periodic table.[10] Linnaeus' fixing of species enabled analogous progress to occur toward a natural system in biology. Importantly, the character of the periodic table suggested underly-ing structure. Scientists investigated regularities across columns and rows. For example, the elements in the rightmost column are gasses at room temperature, and do not react with other elements. Elements in the leftmost column are highly reactive and form compounds with elements in the second-to-rightmost column. Metals cluster across the left two-thirds of the table.

Eventually it was theorized that elements represent discrete combinations of atoms, and that macroproperties of substances result from the orbit and transi-tions of electrons in valence shells. And then it was discovered that elements can in fact transmute. When Marie Curie discovered radium, "element" was explicitly redefined. Though individual samples of elements can be transmuted, the elements remain "fixed" in that they are separated by discrete gaps. There is no "half-uranium, half-thorium" despite the fact that a quantity of uranium can decay into thorium.

Rutherford and Soddy observed atomic transmutation in 1901, and knowledge of the discovery was widespread by 1913.[11] Lotsy (1916, p. 7), in his wide-ranging discourse on "evolution" in different sciences, described how the fixation of chemical elements enabled the formulation of the periodic table and subsequent discovery of "the direct transformability of the elements". While Bateson did not explicitly describe the transmutation of chemical elements, he would have been aware that the concept of element had changed dramatically over time.

Bateson analogized biological species to chemical elements in order to stress the importance of a unit element to scientific progress. The species unit enabled early systematists to conceive a natural system of classification. New research questions made sense in this framework; How many forms of natural relationship are there? What is the nature, origin, and limit of variation within a species? These questions and others spurred 19th century systematic investigation.

Unit characters, unit factors

In addition to biological species, Bateson analogized the units of genetics to chemical elements. Many other early 20th century geneticists also made this analogy (Castle, 1919, p. 127, 1922, p. 8; Davenport, 1906, pp. 78, 80; DeVries, 1909, pp. 3, 330; East, 1912; Frost, 1917, p. 246). What are we to make of the chemistry analogies? Are we to literally conceptualize factors as unit material elements, floating around and interacting in the germ-plasm just as hydrogen and oxygen atoms swirl about in water? In this section, I argue that Bateson, at least, did not intend this materialist interpretation.

Bateson and Saunders[12] (1902) compared the units of heredity to those of chemistry several times in their *Experimental Studies in the Physiology of Heredity*. In speculating as to the reason that Mendel's discovery remained unknown for so long,[13] they noted that Darwin's experiments had failed to shed light on heredity in part because of:

> the habit of regarding various species, breeds, varieties, and casual fluctuations as all comparable expressions of one phenomenon, similar in kind; and to insufficient recognition of the possibility that variation may be, in its essence, specific. . . . Various breeds and various crosses were mixed together, and the results are not unlike those which the early chemists would have arrived at in testing the affinities and constitution of a number of unknown elementary and compound substances mixed together at random.
>
> (Bateson and Saunders, 1902, p. 84)

Bateson and Saunders' point was that the mechanisms of heredity and types of variation were substantially confused. Darwin's "pangens" were not appropriate entities to play the role of theoretical element in genetics, that is, the unit of inheritance. Observed characters were not appropriate to serve as a unit of analysis with respect to investigating heritable morphology.

Bateson and Saunders subsequently explicitly introduced the notion of unit characters of heritable morphology. They described Mendel as having crossed forms exhibiting "antagonistic" characters: the offspring bear exactly one of the observable characters, without evidence of the other character. These unit characters could serve as elements with respect to both morphology and inheritance. The observed unit characters must relate somehow to the germ-plasm. The character units were "the sensible manifestations of physiological units of as yet unknown nature" (Bateson and Saunders, 1902, p. 122).

Does this talk of "physiological units" commit Bateson and Saunders to an atomistic, one-to-one relation of unit-character to material unit-factor as a general rule? Frost described such a view:

> If we adopt a factor-to-factor system of notation, it is natural to conceive of the opposed "factors" as physical units responsible for genetic potentialities. We are thinking of assumed physical units of segregation.
>
> (Frost, 1917, p. 247)

Frost ascribed this view, which I refer to as "materialist", to Morgan (1915, p. 419). Frost identified an alternative:

> [A] factor is not an element of the germ-plasm; it is rather a property or characteristic of the germ-plasm or of some element of the germ-plasm. The characters of an organism, as Gates (1914, p. 269) has remarked, are "attributes", no more to be separated from the organism than are the properties of a chemical compound from that compound. The factors, similarly, are inferred properties or attributes of the germ-plasm, by whose behavior we explain the alternative transmission of certain properties or attributes of the soma. Obviously an organism is not composed of "characters" – and neither is its germ-plasm composed of "factors", . . . like the characters of the soma, [factors] are more or less conventionalized in description. We have no warrant for projecting these conventionalized descriptions back into the actual germ-plasm, and assuming the presence and absence there of strictly corresponding material units of segregation.
>
> (Frost, 1917, p. 246)[14]

The phrase "physiological units of as yet unknown nature" suggests that Bateson and Saunders held the materialist view. When speculating on the possibility of exceptions to the pure transmission of characters, Bateson and Saunders mention the "mosaic" fruits found in *Datura* spp. These exceptional cases could indicate that "the germ-cells may also have been mosaic" (Bateson and Saunders, 1902, p. 92). In turn, this language seems to suggest that "unit factors" are mosaics of different materials. However, Bateson and Saunders explicitly stated that no material basis for the units of heredity should be assumed:

> Remembering that we have no warrant for regarding any hereditary character as depending on a material substance for its transmission, we may, with this proviso, compare a compound character with a double salt, such as an alum, from which one or other of the metals of the base can be dissociated

by suitable means, while the compound acid-radicle may be separated in its entirety, or again be decomposed into its several constituents. Through a crude metaphor, such an illustration may serve to explain the great simplification of the physiology of heredity to which the facts now point.

(Bateson and Saunders, 1902, p. 111)

The context of this last quotation is Bateson and Saunders' discussion of the advantages that Mendelism presents regarding genetic experimentation. The chance that a given individual will transmit a character is "a question to be determined by actual observation"; genetic investigation is "now merely a matter for precise quantitative analysis" (Bateson and Saunders, 1902, p. 111). Bateson and Saunders lamented that Darwin's experiments were too "complex" (read: confused). The metaphor that compares unit factors to unit elements is "crude" in part because Bateson and Saunders are aware that the reader might read a materialist view into it. Hence the explicit proviso pleading agnosticism regarding the material view. However, the metaphor was helpful to illustrate that scientific progress could be made via some idea of the appropriate quantitative units to be investigated. Bateson and Saunders repeat the analogy to emphasize this point:

[N]ot very different from that which opened in chemistry when definiteness began to be perceived in the laws of chemical combination. It is reasonable to infer that a science of Stoechiometry [*sic*] will now be created for living things, a science which shall provide an analysis, and an exact determination of their constituents. The units with which that science must deal, we may speak of, for the present, as character-units, the sensible manifestations of physiological units of as yet unknown nature. As the chemist studies the properties of each chemical substance, so must the properties of organisms be studied and their composition determined.

(Bateson and Saunders, 1902, p. 123)

Let us return to Frost's (1917, pp. 245–246) dichotomy: factors as physical units versus factors as "properties or attributes of the germ-plasm", which he described as "developmental potentialities". Bateson and Saunders (1902) expressed agnosticism about any proposed material basis for the factors. Although in his *Mendel's Principles of Heredity*, Bateson was not specific about the physical nature of the units, he suggested a mechanism for the units' operation:

With the recognition of unit-characters our general conceptions of the structure and properties of living things inevitably undergo a change. . . .What the physical nature of the units may be we cannot yet tell, but the consequences of their presence is in so many instances comparable with the effects produced by ferments, that with some confidence we suspect that the operations of some units are in an essential way carried out by the formation of definite substances acting as ferments.

(Bateson, 1913b, p. 266)

It is scarcely necessary to emphasize the fact that the ferment itself must not be declared to be the factor or thing transmitted, but rather, the power to produce that ferment, or ferment-like body. . . . Next we have to recognize that this antecedent power must be of such a nature that in the cell-divisions of gametogenesis it can be treated as a unit – we have no knowledge as to the actual nature of the factor – and only a conjecture as to whether it is a material substance, or a phenomenon of arrangement.

(Bateson, 1913b, p. 268)

Bateson proceeded to declare that the process of segregation does not appear to be a "process of chemical separation. Its features point rather to mechanical analogies". Bateson's focus on the operation of these units suggests that he indeed conceives unit factors as "developmental potentialities". Throughout *Problems*, Bateson demonstrated a bias against simple, material explanations of characters in favor of mechanistic wave models of development:

When however we pass from the substantive to the meristic characters, the conception that the character depends on the possession by the germ of a particle of a specific material becomes even less plausible. . . . The distinction must surely be of a different order. If we are to look for a physical analogy at all we should rather be led to suppose that these differences in segmental numbers correspond with changes in the amplitude or number of dividing waves than with any change in the substance or material divided.

(Bateson, 1913a, p. 34)

Bateson continued to straddle the fence as regards the nature of factors, to the point that he avoided the term "transmitted": "In attempting to form some conception of the processes by which bodily characteristics are transmitted, or – to avoid that confusing metaphor of 'transmission' – how it comes about that the offspring can grow to resemble its parent?" (Bateson, 1913a, p. 34). The problem with the word "transmit" is that it implies an actual physical transfer of some material unit that is basic to characters. In the materialist view, transmission is not simply a metaphor: material unit factors are actually transmitted from parents to offspring.

Bateson compared Mendel's advance to that made by early chemistry in order to emphasize the lack of knowledge of the material basis of inheritance:

Somewhat as the philosophers of the seventeenth and eighteenth centuries were awaiting both a chemical and a mechanical discovery which should serve as a key to the problems of unorganised matter, so have biologists been awaiting two several clues. In Mendelian analysis we have now, it is true, something comparable with the clue of chemistry, but there is still little prospect of penetrating the obscurity which envelops the mechanical aspect of our phenomena.

(Bateson, 1913a, p. 32)

The "clue of chemistry" is that there are unit elements that do not blend with each other, and which interact in quantitatively fixed ways. This clue says nothing about the physical nature of the unit elements, just as Lavoisier's definition of element said nothing about electrons, neutrons, and protons. Only by recognition of the clue, however, could genetics make progress analogous to that made by early chemists.

This progress required striking down the speculations of alchemists. In a speech to the British Association for the Advancement of Science, Bateson recognized the destructive character of his own research program:

> Somewhat reluctantly, and rather from a sense of duty, I have devoted most of this address to the evolutionary aspects of genetic research. . . . The outcome, as you will have seen, is negative, destroying much that till lately passed for gospel. Destruction may be useful, but it is a low kind of work. We are just about where Boyle was in the seventeenth century. We can dispose of alchemy, but we can not make more than a quasi-chemistry. We are awaiting our Priestley and our Mendeléeff.
>
> (Bateson, 1914, p. 302)

The result of the destruction was to establish research questions and set the track for the development of modern genetics. The subject of genetics would indeed prove to be the title of Richard Goldschmidt's (1940) opus, *The Material Basis of Evolution*. It is likely that Goldschmidt did not intend "material" in (only?) the physical sense. Just as a witness can be material, the emerging theory of genetics was the salient basis for the development of a modern evolutionary synthesis.

Bateson presumed that this salient basis would take the form of a unit of classification at the heritable level, that would account for biological specificity throughout the course of development. This unit of analysis might turn out to be physically instantiated by proteins (Kay, 1993; Olby, 1994), enzymes, or DNA (Deichmann, 2011). Each of these physical entities could be investigated in terms of its ability to produce natural kinds that are "fixed" in the sense of having recognizable, discrete gaps between characters and species. As is demonstrated elsewhere in this volume, the idea of "gappiness" between natural kinds may or may not be present in discussions of synthetic kindhood.

Acknowledgments

I thank Ute Deichmann, two anonymous reviewers, and participants in sessions at the 2018 History of Science Society and Canadian Society for History and Philosophy of Science meetings.

Notes

1 Christine Ladd-Franklin completed all of the requirements for the PhD, including a thesis supervised by Charles Saunders Peirce, in 1883. Johns Hopkins refused to grant

her the PhD until 1927 because she was a woman (Cadwaller and Cadwaller, 1990). Thus I use the term "scholar" rather than "professor", though she studied, published, and eventually was allowed to teach just as any other professor, and was referred to as "an eminent logician" by Bertrand Russell (1948, p. 180).

2 There is a certain historical irony in Bateson's exasperation that the old "lumping" versus "splitting" arguments simply would not go away despite the advance of systematic science. Similar sentiments continue to be expressed more than 100 years later (Garbino et al., 2019). In fact, there is a Wikipedia page on the issue (Lumpers and Splitters, n.d.).

3 Linnaeus allowed that some species might have arisen as hybrids from previously existing species. He insisted that hybridization leading to new species would not occur across genera, however.

4 This interpretation is compatible with the text of Genesis 1. In the New Revised Standard Version: "Then God said, 'Let the earth put forth vegetation: plants yielding seed, and fruit trees of every kind on earth that bear fruit with the seed in it'. And it was so. The earth brought forth vegetation: plants yielding seed of every kind, and trees of every kind bearing fruit with the seed in it. And God said, 'Let the waters bring forth swarms of living creatures, and let birds fly above the earth across the dome of the sky'. And God said, 'Let the earth bring forth living creatures of every kind: cattle and creeping things and wild animals of the earth of every kind'. And it was so".

5 Amundson (2005, n.8) commented that "virtually every philosopher and most historians of biology with whom I have discussed pre-Linnaean transmutationism have been surprised (to the point of incredulity) that species fixism based on essentialism was not the dominant view of species prior to Linnaeus". The work of Amundson, Wilkins (2009), Winsor (2003, 2006), and others has changed the situation among historians and philosophers, at least, though in my experience, biologists continue to be surprised. Bateson (1913a, p. 6) commented that "an interesting piece of biological history might be written respecting the growth and gradual hardening of the conception of Species". Another interesting piece might describe the process by which biological history came to forget that this growth and hardening occurred.

6 Woodruff does not italicize the titles of these monographs, as contemporary convention would. The editor has left the typography as in the original.

7 Huxley proceeded to expound upon the "complete and irreconcilable antagonism" between the theory of evolution and the Catholic Church (the "vigorous and consistent enemy of the highest intellectual, moral, and social life of mankind"). He challenged Mivart to produce, in a third edition of *The Genesis of Species*, actual instances of post-Creation species origin in Suárez's works. In *Lessons from Nature* (Mivart, 1876, pp. 434–435), Mivart did so, including the aforementioned leopard.

8 Indeed, Pouchet (1853, p. 279) claimed that by ordering his natural history of all then-known species alphabetically, the Bishop of Ratisbonne had invented the modern dictionary.

9 "La doctrine de la fixité ne date scientifiquement, quoi qu'on en ait dit, ni du moyen âge ni du xix-ième siècle; ni d'Albert le Grand, ni de Cuvier; mais de Linné". I take his meaning to be that it would be silly to believe that the doctrine originated in the 13th century with Albert, and equally silly to believe that the doctrine originated with Cuvier – whom Blainville famously despised – see Flourens (1865).

10 Chemists of this time may or may not have viewed the periodic table (which wasn't one agreed-upon thing in 1913) as a "natural system". There is a great deal to be said about what "a natural system" even is. I thank Julia Bursten for this point, and I take it to be central to the goals of this volume. (For further discussion, see in particular Chapter 2 and Chapter 7.) Post-Hennigian biologists understand the biological "natural system" in terms of Hennig's (1966) redefinition of phylogeny and related definitions (monophyly, -apomorphy). Systematists and philosophers of systematics (such as myself) are now primarily concerned with the methods for making claims about the agreed-upon biological natural system, phylogeny: the branching pattern of descent

with modification. Hennig analogized the task of phylogenetic systematics to reconstructing a broken pot, hydrological map, or fragments of a map generally, in line with the realization that systematics is a fundamentally historical science: the task is to reconstruct what happened in the evolutionary past.

11 The story is that Soddy shouted "Rutherford, this is transmutation!" to which Rutherford replied, "For Christ's sake, Soddy, don't call it transmutation. They'll have our heads off as alchemists" (Howorth, 1958).

12 Edith Rebecca Saunders graduated from the female-only Newnham College, Cambridge, served as director of the Balfour Biological Laboratory for Women 1890–1914, and received the Banksian Medal from the Royal Horticultural Society in 1906. She was elected to the Linnaean Society in 1905 and to the presidencies of the botanical section of the Royal Horticultural Society in 1920 and the Genetical Society in 1936. See (Richmond, 2001). J.B.S. Haldane (1945) described her as the "mother of genetics" to correct an incomplete obituary published earlier that year in *Nature*.

13 Bateson was one of the three recognized re-discoverers of Mendel's work, together with Hugo DeVries and Carl Correns. J.B.S. Haldane (1945) evidently recognized Elizabeth Rebecca Saunders as a fourth: "It is clear that she and Bateson had independently rediscovered some at least of Mendel's laws before his work was known to them".

14 In his *Experimental Studies in the Physiology of Heredity* (1902, p. 112), and again in *Mendel's Principles of Heredity* (Bateson, 1913b, p. 73), Bateson speculated that a species might in fact be separated from its characters, with a specific "essence" remaining behind.

References

Amundson, R. (2005) *The Changing Role of the Embryo in Evolutionary Thought: Roots of Evo-Devo*. Cambridge: Cambridge University Press.

Bateson, W. (1913a) *Problems of Genetics*. New Haven: Yale University Press.

———. (1913b) *Mendel's Principles of Heredity*. Cambridge: Cambridge University Press.

———. (1914) Address of the President of the British Association for the Advancement of Science. *Science*. 40(1026), 287–302.

Bateson, W. & Saunders, E. R. (1902) Experimental Studies in the Physiology of Heredity. *Reports to the Evolution Committee of the Royal Society*. 1, 1–160.

Bather, F. A. (1927) The Anniversary Address of the President. *Proceedings of the Geological Society*. 83(2), lii–civ.

Blainville, M. & Maupied, F. (1845) *Histoire des Sciences de l'Organisation et de leurs Progrès, comme base de la Philosophie*. Vol. 2. Paris: Librairie Classique de Perisse Frères.

Cadwaller, J. & Cadwaller, T. (1990) Christine Ladd-Franklin (1847–1930). In A. N. O'Connell & N. F. Russo (Eds.), *Women in Psychology: A Bio-Bibliographic Sourcebook* (pp. 220–225). New York: Greenwood Press.

Castle, W. (1919) Piebald Rats and the Theory of Genes. *Proceedings of the National Academy of Sciences*. 5(4), 126–130.

———. (1922) *Genetics and Eugenics: A Textbook for Students of Biology and a Reference Book for Animal and Plant Breeders*. Cambridge, MA: Harvard University Press.

Davenport, C. (1906) *Inheritance in Poultry*. Washington, DC: The Carnegie Institution of Washington.

Deichmann, U. (2011) Early 20-th century research at the interfaces of genetics, development, and evolution: reflections on progress and dead ends. *Developmental Biology*. 357, 3–12.

DeVries, H. (1909) *The Mutation Theory*. Vol. 1. Chicago: The Open Court Publishing Company.

East, E. (1912) The Mendelian Notation as a Description of Physiological Facts. *The American Naturalist*. 46(551), 633–655.

Flourens, J. (1865) Éloge Historique de Marie-Henri Ducrotay de Blainville. *Quarterly Review*. 24(68), 365–380.

Franklin, F. & Franklin, C. (1888) Mill's Natural Kinds. *Mind*. 13(49), 83–85.

Frost, H. (1917) The Different Meanings of the Term "Factor" as Affecting Clearness in Genetic Discussion. *The American Naturalist*. 51(604), 244–250.

Garbino, G., Serrano-Villavicencio, J. & Gutiérrex, E. (2019) What Is a Genus Name? Conceptual and Empirical Issues Preclude the Proposed Recognition of Callibella (Callitrichinae) as a Genus. *Primates*. Online first https://doi.org/10.1007/s10329-019-00714-3.

Goldschmidt, R. (1940) *The Material Basis of Evolution*. New Haven: Yale University Press.

Griffiths, P. (1999) Squaring the Circle: Natural Kinds with Historical Essences. In R Wilson (Ed.), *Species: New Interdisciplinary Essays* (pp. 209–228). Cambridge, MA: MIT Press.

Haldane, J. B. S. (1945) Miss E. R. Saunders. *Nature*. 156(385), 385.

Hennig, W. (1966) *Phylogenetic Systematics*. Translated by D. Dwight Davis and Rainer Zangerl. Urbana, IL: University of Illinois Press.

Howorth, M. (1958) *Pioneer Research on the Atom: Rutherford and Soddy in a Glorious Chapter of Science: The Life Story of Frederick Soddy*. Jacksonville, FL: New World Publications.

Huxley, T. H. (1871) Mr. Darwin's Critics. *Contemporary Review*. 18, 443–476.

Kay, L. E. (1993) *The Molecular Vision of Life*. New York, Oxford University Press.

Locy, W. (1915) *Biology and Its Makers*. 3rd edition. New York: Henry Holt and Company.

Lotsy, J. (1916) *Evolution by Means of Hybridization*. Leiden: M. Nijhoff.

Lumpers and Splitters. (n.d.) *In Wikipedia*. 2018. Retrieved from https://en.wikipedia.org/wiki/Lumpers_and_splitters/ 23 January 2019.

Magnus, P. D. (2014) No Grist for Mill on Natural Kinds. *Journal for the History of Analytic Philosophy*. 2(4), 1–15.

———. (2015) John Stuart Mill on Taxonomy and Natural Kinds. *HOPOS: The Journal of the International Society for the History of Philosophy of Science*. 5(2), 269–280.

Mill, J. S. (1843a) *A System of Logic, Ratiocinative and Inductive: Being a Connected View of the Principles of Evidence and the Methods of Scientific Investigation*. Vol. 1. London: John W. Parker.

———. (1843b) *A System of Logic, Ratiocinative and Inductive: Being a Connected View of the Principles of Evidence and the Methods of Scientific Investigation*. Vol. 2. London: John W. Parker.

Mivart, St. G. (1871) *On the Genesis of Species*. 2nd edition. London: Macmillan and Co.

———. (1876) *Lessons from Nature, as Manifested in Mind and Matter*. London: John Murray.

Morgan, T. H. (1915) The rôle of the environment in the realization of a sex-linked Mendelian character in Drosophila. *The American Naturalist*. 49(583), 385–429.

Olby, R. (1994) *The Path to the Double Helix*. New York: Dover.

Osborn, H. F. (1913) *From the Greeks to Darwin: An Outline of the Development of the Evolution Idea*. London: MacMillan and Company.

Pouchet, F. A. (1853) *Histoire des Sciences Naturelles au Moyen Âge: ou, Albert le Grand et son Époque*. Paris: J. B. Baillière.

Read, C. (1877) On Some Principles of Logic. *Mind*. 2(7), 336–352.

———. (1878) *On the Theory of Logic: An Essay*. London: Kegan Paul & Co.

————. (1898) *Logic: Deductive and Inductive*. London: Grant Richards.

Richmond, M. L. (2001) Women in the Early History of Genetics: William Bateson and the Newnham College Mendelians, 1900–1910. *Isis*. 92(1), 55–90.

Russell, B. (1948) *Human Knowledge: Its Scope and Limits*. New York: Simon and Schuster.

Saint Hilaire, G. I. (1859) *Histoire Naturelle Générale des Règnes Organiques*. Vol. 2. Paris: Librairie de Victor Masson.

Sanson, A. (1900) *L'Espèce et la Race en Biologie Générale*. Paris: Librairie C. Reinwald.

Whewell, W. (1828) *An Essay Concerning Mineralogical Classification: With Tables of the Orders and Species of Minerals*. Cambridge: J. Smith.

————. (1847) *The Philosophy of the Inductive Sciences, Founded on Their History*. Vol. 1. London: John W. Parker.

Wilkins, J. (2009) *Species: A History of the Idea*. Berkeley: University of California Press.

————. (2012) Biological Essentialism and the Tidal Change of Natural Kinds. *Science Education*. 22(2), 221–240.

Winsor, M. P. (2003) Non-Essentialist Methods in Pre-Darwinian Taxonomy. *Biology and Philosophy*. 18, 387–400.

————. (2006) The Creation of the Essentialism Story. *History and Philosophy of the Life Sciences*. 28, 149–174.

Woodruff, L. L. (1921) History of Biology. *The Scientific Monthly*. 12(3), 253–281.

A new synthesis of concerns about biological kinds

4   Artifacts and artefacts
A methodological classification of context-specific regularities

Vadim Keyser

Introduction

The ontological classification of scientific objects can be informed by the epistemological methodology ("methodology" hereafter) of differentiating those objects. Proper methods can serve as *indicators* for how to classify scientific objects and processes (Keyser, 2016a, 2016b, 2017). One such method involves using multiple identifications, measurements, and/or experiments to differentiate genuine phenomena from results produced in error. When independent methods converge on an object or process, inferences are made about its status as a natural phenomenon rather than, for example, a systematic blunder attributable to the measurement setup. Two examples that I have previously discussed are Avogadro's number (Keyser, 2016b) and arsenic-consuming bacteria (Keyser, 2016a, 2017). Perrin's (1913) multiple methods for identifying Avogadro's number support its objectivity as a natural phenomenon. Unsuccessful convergence serves as an indicator for error. There has been scientific disagreement about the so-called arsenic-consuming organism, studied by Wolfe-Simon et al. (2010). In 2010, a fundamental issue in biology was in question: whether cellular life requires phosphorous. A bacterium was found to survive (and perhaps thrive) in the waters of Mono Lake, which are filled with arsenic filled. In the laboratory, this bacterium appeared to replace phosphorous with arsenic in its DNA backbone (Wolfe-Simon et al., 2011). But by using more stringent *independent* experimental methods – for example, to remove any trace of phosphorous and to "wash" the clinging arsenate from the DNA samples – Reaves et al. (2012) did not find the DNA structure to have covalently bound arsenate in the DNA structure. Disagreement in results between the two experiments serves as an indicator that the original arsenic-consumption effect was an *artifact* of the lack of purification in the preparatory procedure (Keyser, 2012).

"Robustness analysis" refers to the aforementioned method of multimodal measurement/experiment. Scientists use robustness analysis by comparing multiple independent measurements, experiments, derivations, and models to see if they converge on a given result (Wimsatt, 2007, p. 43).[1] In Wimsatt's characterization, robustness analysis distinguishes the ontologically and epistemologically "trustworthy" from what is unstable and thus unreliable, and it grounds realism, reliability, and objectivity (p. 56). Relevant to my point about the relation between ontology and methodology, robustness analysis can be used to differentiate real objects and processes from "artifacts", or, results produced in error (p. 38).[2] According to Wimsatt, "artifactual" results are unstable across multiple independent experiments and measurements (p. 56).

A signal detection analogy is useful to describe how robustness analysis picks out genuine phenomena and cancels out the noise. Unstable details between independent methods fade into the background, but the robust property or consequence pierces through. Wimsatt (2007) discusses the example of using multiple independent imaging techniques to detect properties of planets. Each imaging technique will contain strong noise, but the signal is weak, so how does it pierce through? By combining techniques, the different types of noise will become weak because they are particularly situated within each technique, but the signal that is consistent over *all* of the techniques will increase in strength (p. 57). The signal is invariant across all methods, but the different types of noise are random and independent and differ between each technique.

Much attention has been paid to using robustness analysis *to differentiate natural phenomena from results produced in error* (Levins, 1966; Weisberg, 2006; Wimsatt, 2007; Lloyd and Parker, 2009; Lloyd, 2010). In Keyser (2016a), I argue that there are regularities other than natural phenomena that require careful philosophical attention. Experimental "effects" are produced by the manipulation of experimental conditions, and they are difficult to classify alongside results produced in error ("artifacts") as well as "natural phenomena".

Hacking (1983, p. 221) distinguishes "effects" from "phenomena", which are observable regularities that are not the result of experimental intervention. If Hacking is right, then few phenomena in nature are "waiting" to be observed; however, science is populated with effects (p. 227). While effects may be plentiful, they are condition dependent and thus fall apart when specific conditions in a given experimental setup are not met, so one challenge in scientific practice is to be able to create and re-create effects. Hacking summarizes the "aim of experiments" as creating, refining, and repeating effects produced in a given experiment (pp. 229–230). Elsewhere, I argue that if we apply robustness analysis to a given effect, we *cannot figure out if it is reliable or unreliable* (Keyser, 2016a). But there is a further point, which will be the focus of this discussion. If we apply robustness analysis to a given effect, we *cannot figure out if it is a genuine phenomenon or an artifact*. That is, even if scientists develop independent methods for the production of some effect, because effects are sensitive to conditions, convergence would likely be unsuccessful, and the effect would fail to be stable. On Wimsatt's (2007) view, this would imply that the effect is an artifact. Without a *methodological* indicator for the reliability, genuineness, and objectivity of a given effect, its classification is uncertain. An important question arises that is both methodological and ontological: *how do we classify context-sensitive scientific productions?* In this case, the methodology of robustness analysis fails to give us proper classificatory guidance. So we must either find a new methodology to guide our ontological classification or separate the ontology from the methodology. In this discussion, I take the former route. I believe that ontological classificatory schemes benefit from the success and failure of applied epistemological methodology. The codevelopment of ontological classification alongside methodology is important for philosophy of science work that speaks to the applied sciences. It is the failure of robustness analysis that brings us to our current ontological question. It will also be the success of a different type of robustness methodology that guides us to an ontological answer.

In this discussion, I argue that context-sensitive effects require a new ontological category: "artefact". These are context-specific/context-sensitive regularities, which are scientifically valuable for some given purpose. I argue that to *differentiate* an artifact from an artefact requires understanding *how* changes in conditions alter given effects, but such differentiation also requires theoretical, experimental, and pragmatic considerations.

The discussion will be outlined as follows. In the next section, I discuss the difficulty of differentiating effects from artifacts. Following that, I present artefactual regularities and discuss the fluid relationship between artifact and artefact. I apply this classification to two empirical puzzles.

Differentiating effects from experimental artifacts

Both effects and experimental artifacts are condition sensitive and, for this reason, difficult to distinguish. I discuss the ontological and methodological implications.[3] First, I characterize "effect" and "artifact" by drawing on an argument from Kroes (2003). Second, I discuss the difficulty of distinguishing effects from artifacts.

Characterizing "effect" and "artifact"

Hacking distinguishes "effects" from "phenomena":

> Phenomena and effects are in the same line of business: noteworthy discernible regularities. The words "phenomena" and "effect" can often serve as synonyms, yet they point in different directions. Phenomena remind us, in that semiconscious repository of language, of events that can be recorded by the gifted observer who does not intervene in the world by who watches the stars. Effects remind us of the great experiments after whom, in general, we name the effects: the men and women, the Compton and Curie, who intervened in the course of nature, to create a regularity which, at least at first, can be seen as regular (or anomalous) only against the further background of theory.
>
> (Hacking, 1983, pp. 224–225)

Effects are condition-sensitive – that is, sensitive to how conditions are arranged. We can apply this to the ordinary Hall effect, the appearance of a potential difference across the conductor at right angles to the magnetic field and the current of the conductor. According to Hacking, the Hall effect occurs only in a particular type of "arrangement" where electrical conductors and magnetic fields are located in just-so relations to one another (Hacking, 1983, p. 224). The original apparatus used to demonstrate the Hall effect was human-made and the conditions were carefully planned, but we have the intuition that the phenomenon was "discovered" in the lab rather than "created" in the lab (p. 225). On Hacking's view, the arrangement of conditions behind the Hall effect occurs *only* in the laboratory:

> I suggest, in contrast, that the Hall effect does not exist outside of certain kinds of apparatus. Its modern equivalent has become technology, reliable and routinely produced. The effect, at least in a pure state, *can only be embodied by such devices.*
>
> (Hacking, 1983, p. 225. My emphasis.)

Hacking's mention of technology and the embodiment of effects in devices is interesting because both experimental and technological effects are sensitive to *the manipulations of the arrangement of conditions* (Keyser, 2016a). The same goes for artifacts. So is the manipulation of conditions the only factor that characterizes "effects"? Alone, this factor is not helpful in differentiating effects from artifacts.

While there are many relevant philosophical views of artifacts (e.g., Simon, 1981; Rasmussen, 2001; Chakrabarty, 2012), Kroes (2003) extends Hacking's discussion and makes a fine-grained characterization of experimental artifacts by discussing "artificial environments" versus the "object system" of study. Drawing on Franklin's (1986) discussion, he characterizes experimental artifacts as "results that are generated by the artificial environment or artificial means of observation of the natural phenomena under study" (Kroes, 2003, p. 71). Kroes writes:

> The results of an experiment are always the outcome of the object system interacting with an artificial environment, and therefore it is always necessary to filter out the component in the results that tells us something about the object system.
>
> (Kroes, 2003, p. 71)

While this ontological and methodological suggestion is useful in distinguishing phenomena from artifacts, it is not helpful for distinguishing effects from artifacts. The implicit assumption here is that we can filter out the artificial environment from the object system in a given effect. However, experimental arrangements that constitute a given effect rely on the manipulation of conditions, where the distinction between "natural" and "artificial" is not only unclear but also indeterminate. For example, Hacking (1983, p. 226) says, "If anywhere in nature there is such an arrangement, with no intervening causes, then the Hall effect occurs". On Hacking's view, a human-made effect can occur in nature, given that the arrangement occurs. Likewise, what is natural can be re-created in a laboratory as a mimetic production (van Fraassen, 2008).

I take the view that it is not the case that some arrangements are *intrinsically* natural and others are artificial, but rather some arrangements are *used* as a natural phenomenon and others are used as artificial *for the specific aims of the experiment, based on theoretical and pragmatic considerations*. In the next section, I argue that what is relevant to the classification of an arrangement is *how* the conditions are *arranged*, and how the experimental practice is *used* by scientists.

Methodologically differentiating effects and artifacts

Part of the suggestion presented by Kroes (2003) is methodological. It is "necessary" to filter out the object system from the artificial environment to obtain proper results. Here, we return to robustness analysis, which is designed to filter out error and noise. But as we saw earlier, robustness analysis helps only in the context of discovered natural phenomena. The mechanism for the success of robustness analysis is that condition sensitivity serves as the indicator that a result is produced in error. In the context of "arsenic-consuming" bacteria, robustness analysis works well.

At first, it seems as though scientists have measured carefully by checking for bound arsenate (Wolfe-Simon et al., 2011). But as Reaves et al. (2012) show, stricter standards for "washing" the DNA structure results in no covalently bound arsenate in the DNA backbone, which implies that the bacterium does not replace phosphorous with arsenate in its DNA structure. Using varying and multiple methods, Reaves et al. (2012) make the argument that the original pre-spectroscopy DNA "purification" procedure resulted in error. The experimental progression from Wolfe-Simon et al. (2011) to Reaves et al. (2012) can be interpreted as making methods finer grained to give precise and accurate information about DNA-binding – answering questions like, is the arsenate tightly bound, or is it merely lingering on the backbone? In this context, the binding (or failure of binding) of arsenate to DNA is *independent of the arrangement of conditions in the preparatory procedures*. If the preparatory procedure (e.g., the washing procedure) is not properly performed, error clouds the appearance of binding (or failure of binding).

In contrast to the aforementioned example, in the case of effects, the thing produced is *arrangement-dependent*. Take an example of a currently unraveling case of measurement in ecological-developmental measures in reptiles. In all crocodiles and in the majority of turtles, along with a few species of lizards, sex determination is environmental. This means that certain environmental features determine the sex of offspring. A particular version of environmental sex determination is temperature-dependent sex determination (TSD). In TSD, the temperature during a specific developmental window in incubation determines the sex of an organism. As discussed in Keyser (2012), TSD can be neatly summarized into four components:

1   Under TSD, the sex of individuals is determined permanently after conception, by incubation temperature.

2   If temperature is above a pivotal temperature, it will produce one sex, and if it is below the pivotal temperature, it will produce another. Sometimes two pivotal temperatures exist.
3   Three patterns of sex determination have been identified:

> *TSDIa*: Male-Female (low temperatures produce males, and high temperatures produce females).
> *TSDIb*: Female-Male (low temperatures produce females, and high temperatures produce males).
> *TSDII*: Female-Male-Female (low temperatures produce females, intermediate produce males, high produce females).

4   The effect of temperature on sex determination is exerted during a small window of time during development called the thermosensitivity period (TSP).

Of particular interest to us is *how* effects and artifacts occur in TSD measurement in the field and lab. Measuring developmental dynamics in TSD *seems* simple: scientists need to isolate the natural from the artificial. It requires that scientists properly intervene in the organism to measure certain developmental phenomena. Some examples of developmental phenomena of interest are the TSP as it is located within the developmental timeline, the pattern of sex determination, and even the flip-over response (righting response) (Valenzuela, 2001, 2004; Paitz et al., 2010). Scientists have found that developmental results differ between lab and field conditions. Specifically, field measurement setups change the starting time and duration of the TSP (the period at which the organism is sensitive to temperature effects). The specific reason for this is because temperature fluctuations affect developmental rate by accelerating or decelerating development via enzymatic activity. Changes in enzymatic activity cause TSP to begin at different times and last for different lengths of time (Keyser, 2012). The causal roles of measurement conditions, like pulses of cool temperatures, are still unclear (Valenzuela, 2004). However, I interpret these results to show that developmental "phenomena" like TSP are dependent on conditions in a given experimental or measurement setup (Keyser, 2012). In the laboratory, scientists control temperature conditions to be constant temperature conditions, but in semi-natural and natural setups, scientists have to bury eggs, control the moisture of the soil, and perform invasive or noninvasive measures of development (Bull, 2004).

It is important to note that the history of TSD measurement in "natural conditions" involves manipulation: Bull describes his initial field work as involving burying eggs in the coldest possible nesting sites and purchasing black aquarium gravel in order to replace the sand over sun-exposed sites (Bull, 2004, p. 6). Both field and lab conditions require the arrangement of conditions. The "phenomena" that result from the arrangements are context-sensitive objects of study.

It is not just TSP that is context-sensitive, but also immune response, size mass, and righting response (Paitz et al., 2010). Elsewhere (Keyser, 2012), I argue that both the laboratory and the field setups are instructive for understanding how the specific causal conditions work in TSD in reptiles. By focusing on the conditions within both lab and field conditions, we get more *causal information* about specific factors – for example, the effect of temperature on aromatase – as well as larger "phenomena" such as fitness values that differ between lab and field setups.

Of particular importance to this discussion is that TSD measurement/experiment *seems* like a cut-and-dry case of filtering out the natural from the artificial. However, what is "natural" requires manipulations made by scientists in the form of controlling soil moisture/substance and depth of egg burying. Classifying the "phenomena" from this type of experiment as "natural" would be idealizing the picture. The developmental results are condition-sensitive, and scientists are *causally intervening* to control those conditions. Likewise, what is "artificial" is instructive about fine-grained causal connections such as the role of aromatase during TSP. Classifying TSD lab measurements as artificial and thus artifactual overlooks their causal importance. A more adequate classification of TSD measurement is in terms of how the specific developmental measures are *produced*. This would avoid setting arbitrary or idealized boundaries between the artificial and natural. It would also avoid the methodological impossibility of filtering out the artificial. I discuss this in the next section.

When it comes to effects, I propose that the failure to filter out the artificial from the natural in a measurement/experimental setting is not merely a failure of classification due to epistemic limits. The way in which effects are arrangement-dependent makes it impossible to filter out the "natural" from the "artificial". Given that artifacts are arrangement-dependent also, we need another condition that differentiates the two. I propose that to differentiate an artifact from an artefact, ontologically and methodologically, requires understanding *how* changes in conditions alter given effects; however, such differentiation also requires theoretical, experimental, and pragmatic considerations.

Artifacts and artefacts

Earlier, I referenced robustness analysis to argue flaws in classifying scientific objects using methodology. But as I argue elsewhere (Keyser, 2016b), robustness analysis can be usefully reframed to focus on *causally tracking* measurement results. In "causal tracking", the methodological and ontological lenses shift from focusing on some thing as *either* natural *or* artificial to focusing on *tracking the conditions or processes relevant to the artefact of study*. That is, the focus is on *how a given context-sensitive object or process varies with changes in its arrangement*. However, there are important pragmatic, experimental, and theoretical/modeling considerations to be made about any artefact of study. For example, what kind of result counts as "problematic" for the *purposes* of some experiment? I discuss these considerations ahead. I define "artefact" as a context-sensitive object of study that has experimental, theoretical, and/or pragmatic scientific value.

Causally tracking artefacts of study

Experiments with context-sensitive arrangements are useful for understanding *how* specific experimental conditions *produce* differences and similarities in the object of study. Considering this methodological point re-focuses our ontological lens. Instead of focusing on the distinction between the natural and the artificial, I suggest that it is philosophically important to analyze the *production process*. The production process will contain experimental conditions (i.e., "arrangements") that are causally relevant to the production of a given artefact.[4] In Keyser (2016b), I spell out how conditions in multiple production processes can be compared in order to causally track what is unfolding in an experiment. This process requires a comparison of converging and diverging *conditions* in each production process, as well as converging and diverging *results* of each production process (Keyser, 2016b). For our purposes, the outcome of this two-level comparison can be used to explain how certain conditions produce (or fail to produce) specific changes in a given artefact. That is, analyzing the production processes informs *how* conditions change artefacts.

In Keyser (2016b), I assume an important methodology that is attributed to Woodward (2003). If one variable is manipulated and this results in changes to another variable, there is an indicator of a causal relation. To make the methodology of manipulation causally informative, I argue that once experimental conditions are compared and results are compared, *theory* can be used to explain *why* converging and diverging results occur. A simple example is in boiling point production (Keyser, 2016a). We know from Chang (2004) that the boiling point of water depends on specific conditions, such as atmospheric pressure and dissolved gas. Differences in both conditions result in variation in the boiling point. So boiling point is a context-sensitive object of study – artefact. Scientists can run multiple experiments where conditions are altered (e.g., no dissolved gas vs. presence of dissolved gas) to see if there is convergence or divergence of results. Once scientists observe that there is divergence in results based on specific manipulations in conditions, they can use theory to home in on the particular relations between atmospheric pressure, dissolved gas, and heat. Without theoretical application, there is a failure in *characterizing* the causal story behind the artefact. To summarize, converging and diverging results point out *which* conditions produce a change in the artefact. That is, the cross-comparison in conditions *locates specific changes* in the production processes (Keyser, 2016a). Theory is used to characterize how or why the conditions are relevant to the production process.

Applying this method to the case of TSD results in a useful ontology for understanding developmental artefacts. Lab, semi-natural, and even computer simulation experiments all require manipulations of conditions ("arrangements"). For example, in lab experiments, temperatures are traditionally constant rather than fluctuating. If we were to heed the advice of some biologists, we would discard lab setup results. For example, Delmas et al. (2007) call the lab setups "proxies" and claim that the use of proxies has been "refuted" in the TSD literature. Notably, the studies they cite merely show a difference between lab and field results rather than a "refutation" of lab measures. The suggestion to discard lab results assumes a view that field setups are ontologically primary and methodologically superior. However, both lab and field conditions allow for tracking specific *variant and invariant results*. That is, certain kinds of conditions in lab vs. field produce relevant differences in results, while other condition changes do not produce differences in results.

Some examples of variant results are:

1 Embryos incubated at constant temperatures in the lab attain more advanced stages of development every week than any of the field treatments.
2 Embryos from sunny nests developed faster than embryos in shaded nests and covered experimental settings.

(Keyser, 2012)

Invariant results do not change with changes in measurement conditions. For example, development is faster at higher constant incubation temperatures and higher mean temperatures for both the lab and field (Keyser, 2012).

If certain measurement/experimental setups are discarded, then these comparisons, which are useful for further exploring the causes of variant or invariant information, become lost. This is precisely the problem with ontological focus on the natural vs. the artificial rather than the production process. If the ontological focus shifts to the production process, then both field and lab conditions are reduced to arrangements (e.g., constant temperature vs. temperature pulse arrangements) that are *causally relevant to certain artefacts* (e.g., slowed righting response vs. faster righting response).

A possible criticism here is that perhaps not all artefacts deserve the same ontological status. For example, suppose that when measuring a specific hormone in turtles, we use an invasive method, which is predicted to have no effect on development. But contrary to prediction, the invasive method disrupts later turtle pheromone effects (e.g., hatchlings can influence other hatchlings through pheromone communication). Surely we would argue that this disruption is no artefact (context-sensitive experimentally useful result), but rather an artifact (a result produced in error). Furthermore, such an artifact of the invasive method should be discarded to avoid further errors. But it can be counter argued that the invasive technique and the resulting effect are experimentally useful because we learn a specific causal relation between our intervention and a given pheromone communication system. This way of characterizing the experiment leaves the artefact out of artifactual territory.[5] This criticism points out the difficulty of drawing the boundary between artefacts and artifacts. It is unclear how the boundary can be drawn with *only physical causal considerations*. We are back to our original problem of distinguishing artifacts from effects as well as error-prone from reliable objects of study. There are important reasons to characterize some things as error-prone and context-sensitive and other things as reliable and context-sensitive. I suggest that the way around these worries is to go beyond physical causal considerations and to look at theoretical, experimental, and pragmatic considerations.

Theoretical and pragmatic considerations

To understand the intersection between classification and methodological causal analysis, I will use the example of measuring the mesosome. The complex story of the mesosome can be framed as story about methodology and ontology.[6] At first, the mesosome seemed to be a natural phenomenon – a structure that is part of the cell and that has a given function. Interestingly, initially the mesosome appeared to be *invariant* under different measurement techniques – that is, it could be detected under different types of microscopes. However, given more precise methods, the mesosome was re-classified as a product of the chemical fixation in a particular type of preparatory procedure. In Wimsatt's (2007) reference to Culp's (1994) discussion, the mesosome is classified as a *product* of preparation methods. The classification codeveloped with precise experimental methods. Scientists modified preparatory conditions to observe how the mesosome is produced – that is, *under what causal conditions* it is produced. Wimsatt nicely summarizes the development of method as a "recipe book for how to produce or avoid mesosomes" (Wimsatt, 2007, p. 381). Notice that the methodological modifications run side-by-side with changes in ontological classification. The mesosome was causally stable under different microscopes, and for this reason was classified as an invariant experimental object. As mentioned earlier, invariance serves as an indicator that the object at hand is a natural phenomenon. The classification of the mesosome as a genuine component in the cellular structure inferentially follows. However, as experimental methods developed, the mesosome was observed to be variant due to the causal conditions in different preparatory procedures. Ebersold et al. (1981) found that when cryofixation followed by "freeze-substitution"[7] procedures were used, the mesosome did not appear. This failure of production under different preparatory procedures changes the classification of the mesosome to a product of the preparatory procedure. Currently, the mesosome is characterized as an invagination in the cytoplasmic membrane. But the characterization is bound to an explanation of the methodology: the invagination in the cytoplasmic membrane is produced by preparation-induced contractions of the nucleosome, due to cytoplasmic membrane damage (Wimsatt, 2007, p. 381). This example is informative because it shows *how* methodology codevelops with the classification of an object of study.

I propose that *characterizing an object of study requires theoretical, experimental, and pragmatic considerations that codevelop with causal methodological considerations*. My account draws from the structure of van Fraassen's (2008) detailed description about the coevolution between measurement procedures and theoretical characterization. According to van Fraassen, scientists use a given theory to classify what is investigated in addition to the procedures used for investigation (p. 124). Van Fraassen's description is important because it shows the close relation between theory and applied practice: as a scientific practice develops, theory is used to *classify what is measured* (p. 139). For example, in the case of measuring temperature, scientific measurement techniques are replaced along with their theoretical underpinnings. One instance is that liquid thermometers were replaced by the gas thermometer, which developed alongside Boyle's gas law and the classificatory system provided by kinetic theory (van Fraassen, 2008; Chang, 2004).

I agree with general points made by van Fraassen in relation to the joint evolution of theory and practice. But it is important to qualify that some scientific contexts do not contain fully developed theory, but rather theoretical, experimental, and pragmatic considerations that are underdeveloped but still *informative* for classification purposes. For example, in the case study of TSD measurement, at one point the role of androgens was unclear. But there was an operational understanding of relationships that occurred when androgens were introduced and when aromatase was inhibited. For example, it was observed that different combinations of aromatase-inhibition and testosterone produced different patterns of sex determination (Rhen and Lang, 1994). While the causal patterns in the experiment unfolded, scientists did not have a theoretical understanding of the role of androgens as the signal for male differentiation. My suggestion is that to be useful for differentiating objects of study, classification can come from theory, operational/experimental considerations (like the one just discussed), and/or pragmatic considerations.

I take the view that *classifying* an object of study involves *characterizing* it in terms of parameters.[8] Traditionally, there is a difference between "parameter" and "variable". "Parameters" refer to universal or invariant constants in a given modeling setup. "Variable" defines system characteristics within a modeling setup but with changing values. Variables can become parameters in certain modeling setups. For the sake of simplicity, I refer to "parameters" as *representational components that can be used to characterize an object of study*. Characterizing an object of study using parameters also involves specifying relations between those parameters – for example, by using a mathematical formulation to represent an empirical regularity. For instance, $PV = nRT$ informs our characterization of temperature in terms of the relation between volume and pressure.

An important point to note is how the results of a given production are characterized and thus classified depends on what kind of parameters and relations are *used* by scientists to *interpret* information about the object of study. Parameters can be provided by (1) well-developed theory, (2) experimental content about the relation between our instruments and some object of study, and/or (3) practical considerations relevant to the experiment. (1)-(3) determine if something is error-prone and context-sensitive or reliable and context-sensitive. For example, the mesosome is interpreted as an invagination in the cytoplasmic membrane. But its classification as an artifact or artefact will depend on what kind of scientific purpose is at hand, which is based on experimental and practical considerations. If the scientific experiment is about functions of genuine cellular components, then it is useful to classify the mesosome as an artifact. But if the experiment is to understand how contractions in the nucleosome produce cytoplasmic membrane damage and mesosome-like structures, the mesosome can be classified as an artefact. The *context of the study* determines the classification.

I take the line between artifacts and artefacts to be fluid because theoretical, experimental, and pragmatic considerations vary. For example, in TSD, measuring what counts as a *relevant temperature profile* can change depending on certain assumptions. In my earlier discussion of TSD, I mentioned a comment by Delmas et al. (2007) that constant temperature setups are just "proxies" and should be replaced. This assumption prompted them to engage a new type of measurement to compare to field setups: computer simulation setups that mimic field setups.[9] They assume that only field setups provide the *relevant temperature profile*. This would mean that any developmental result produced under constant temperature setups is an artifact. When making claims about what will happen to turtle populations, it may be important to take seriously the assumption made by Delmas et al. (2007). With rising mean temperatures, we should expect to see larger thermal fluctuations (Valenzuela, 2004), so it seems that lab results will be artifactual because those results are based on constant temperatures. However, if we take into account more considerations, our classification of lab results shifts again: turtles also make egg-burying choices when it comes to responding to fluctuating temperatures, which can result in turtles burying eggs deeper where temperatures are more constant.[10] Taking this consideration into account, it may be useful to study constant temperature setups because as climate fluctuates, it will affect burial choice and depth in turtles. So lab setups can be considered artefact-producing. Similarly, field conditions may have different temperature profiles than those in truly "wild" conditions. It is not the case that the measures used in the field and also in computer simulations are free from any experimenter manipulation. As mentioned earlier, Bull discussed manipulating burial depth, sun, and sand conditions in field measures. But it would be inaccurate to classify a particular setup (in Bull's case, a field setup with manipulations) as artifact-producing given that the results generated tell us something about the causal relationship between temperature profiles and for example, developmental stage shifts. By shifting considerations, we shift what counts as an artifact vs. an artefact.

Sometimes the boundaries of an experiment are strictly outlined to differentiate what is pre-instrumental from what is produced by instrumental intervention that causally influences a *natural phenomenon*. In such a case it would seem that what is artifactual is tightly determined. But often, practical considerations are important even in contexts with natural phenomena. For example, blood pressure measurements are context-sensitive due to stressful conditions (e.g., being in a hospital setting, anticipating health related outcomes). But because of pragmatic considerations, we do not classify blood pressure measurements as always being artifactual. Some blood pressure readings can be classified as artifactual, but this is due to other considerations from theory and practice that tell us about the relationship between heightened states and effects on blood pressure. While it is outside of the boundaries of this discussion, elsewhere I make the point that natural phenomena are not free from context-sensitive conditions (Keyser, 2012). This makes the line more fluid between "natural phenomena" and "artefacts", as well as between "artefacts" and "artifacts".

Concluding remarks

In this discussion I have argued that context-sensitive effects require a new ontological category: "art*e*fact". I argue that to *differentiate* an artifact from an artefact requires understanding *how* changes in conditions alter given effects; however, such differentiation also requires theoretical, experimental, and pragmatic considerations. A proper extension of this discussion requires a technical account delineating how tracking production conditions is informative about the production of artefacts – the starting steps for which are in Keyser (2017) – and how characterization and classification works between the different sciences. Such a discussion would place an emphasis on the role of *manipulating conditions* to track artefacts and would explore if the production of artefacts is the same in all of the sciences.

Notes

1 For a closer look at robustness analysis in evidence and experimentation, see Horwich (1982), Hacking (1983), Franklin (1997), Sober (1989), Cartwright (1991), Trout (1998), Culp (1994), Woodward (2006), and Stegenga (2009). For robustness analysis modeling, see Levins (1966), Glymour (1980), Weisberg (2006), and Wimsatt (2007).
2 In this discussion, I refer to "artifact" within the boundaries of experiment. In Keyser (2017), I discuss the relation between experimental and technological artifacts.
3 In Keyser (2016a), I discuss the epistemological implications of condition sensitivity for modifying robustness analysis.
4 I am not committed to a particular theory of causation.
5 As pointed out by a reviewer, by reframing the issue yet again (e.g., by looking at the species level vs. individual level), an artefact can seem like an artifact. This is precisely why we need something beyond physical causal considerations to differentiate the two.
6 There is debate about the details of the mesosome story. See (Culp, 1994; Rasmussen, 1993, 2001; Hudson, 1999). I present a picture that focuses on the characterization and the staining procedure methodology.
7 That is, substituting ice with an organic solvent containing the fixative.
8 I spell this view out in detail in Keyser (2012).
9 Important to note is that such a comparison would fail to give true robustness because the two setups are not independent in underlying conditions.
10 See (Keyser, 2012) for a discussion of turtles and measurement setups: the manipulation of setups also occurs by turtles – not just scientists.

References

Bull, J. J. (2004) Perspectives on Sex Determination: Past and Future. In N. Valenzuela & V. Lance (Eds.), *Temperature-Dependent Sex Determination in Vertebrates*. Washington: Smithsonian Books.

Cartwright, N. (1991) Replicability, Reproducibility, and Robustness: Comments on Harry Collins. *History of Political Economy*. 23, 143–155.

Chakrabarty, M. (2012) Popper's Contribution to the Philosophical Study of Artifacts. In Philosophy of Science Assoc. 23rd Biennial Meeting (San Diego, CA) > PSA 2012 Contributed Papers.

Chang, H. (2004) *Inventing Temperature: Measurement and Scientific Progress*. Oxford: Oxford University Press.

Culp, S. (1994) Defending Robustness: The Bacterial Mesosome as a Test Case. In D. Hull, M. Forbes & R. Burian (Eds.), *PSA 1994* (pp. 47–57). East Lansing: Philosophy of Science Association.

Delmas, V., Baudry, E., Girondot, M. & Prévot-Julliard, A. C. (2007) The Righting Response as a Fitness Index in Freshwater Turtles. *Biological Journal of the Linnean Society*. 91, 99–109.

Ebersold, H. R., Cordier, J. L., & Lüthy, P. (1981). Bacterial mesosomes: Method dependent artifacts. *Archives of Microbiology*, 130(1), 19–22.

Franklin, A. (1986) *The Neglect of Experiment*. Cambridge, MA: MIT Press.

———. (1997) Calibration. *Perspectives on Science*. 5, 31–80.

Glymour, C. (1980) *Theory and Evidence*. Princeton, NJ: Princeton University Press.

Hacking, I. (1983) *Representing and Intervening*. Cambridge: Cambridge University Press.

Horwich, P. (1982) *Probability and Evidence*. Cambridge: Cambridge University Press.

Hudson, R. G. (1999) Mesosomes: A Study in the Nature of Experimental Reasoning. *Philosophy of Science*. 66, 289–309.

Keyser, V. (2012) *Measuring Biological Reality: The Stabilization View of Measurement*. (Doctoral dissertation).

———. (2016a) Effects and Artifacts: Robustness Analysis and the Production Process. In Philosophy of Science Assoc. 25th Biennial Mtg (Atlanta, GA) > PSA 2016 Contributed Papers.

———. (2016b) A New Theory of Robust Measurement. American Philosophical Association 2016 Pacific Division Meeting. Retrieved from www.apaonline.org/members/group_content_view.asp?group=110424&id=476093.

———. (2017) Experimental Effects and Causal Representations. *Synthese*, SI: Modeling and Representation, 1–32. https://doi.org/10.1007/s11229-017-1633-3.

Kroes, P. (2003) Physics, Experiment, and the Concept of Nature. In H. Radder (Ed.), *The Philosophy of Scientific Experimentation*. Pittsburgh, PA: University of Pittsburgh Press.

Levins, R. (1966) The Strategy of Model Building in Population Biology. *American Scientist*. 54, 421–431.

Lloyd, E. A. (2010) Confirmation and Robustness of Climate Models. *Philosophy of Science*. 77, 971–984.

Lloyd, E. A. & Parker, W. (2009) Varieties of Support and Confirmation of Climate Models. *Proceedings of the Aristotelian Society*. 83, 1467–8349.

Paitz, R. T., Gould, A. C., Holgersson, M. C. N. & Bowden, R. M. (2010) Temperature, Phenotype, and the Evolution of Temperature-Dependent Sex Determination: How Do Natural Incubations Compare to Laboratory Incubations? *Journal of Experimental Zoology Part B: Molecular and Developmental Evolution*. 314B, 86–93.

Perrin, J. (1913) *Les Atomes*. Paris: F. Alcan. (*Atoms* by D. Li. Hammick, Trans.). New York: D. Van Nostrand. Reprinted Kessinger Publishing, 2007.

Rasmussen, N. (1993) Facts, Artifacts, and Mesosomes: Practicing Epistemology with the Electron Microscope. *Studies in History and Philosophy of Science Part A*. 24, 221–265.

———. (2001) Evolving Scientific Epistemologies and the Artifacts of Empirical Philosophy of Science: A Reply Concerning Mesosomes. *Biology and Philosophy*. 16, 629–654.

Reaves, M. L., Sinha, S., Rabinowitz, J. D., Kruglyak, L. & Redfield, R. J. (2012) Absence of Detectable Arsenate in DNA from Arsenate-Grown GFAJ-1 Cells. *Science*. 337(6093), 470–473.

Rhen, T. & Lang, J. W. (1994) Temperature-Dependent Sex Determination in the Snapping Turtle: Manipulation of the Embryonic Sex Steroid Environment. *General and Comparative Endocrinology*. 96(2), 243–254.

Simon, H. (1981) *The Science of the Artificial*. Cambridge: MIT Press.

Sober, E. (1989) Independent Evidence about a Common Cause. *Philosophy of Science*. 56, 275–287.

Stegenga, J. (2009) Robustness, Discordance, and Relevance. *Philosophy of Science*. 76, 650–661.

Trout, J. D. (1998) *Measuring the Intentional World*. Oxford: Oxford University Press.

Valenzuela, N. (2001) Constant, Shift, and Natural Temperature Effects on Sex Determination in Podocnemis Expansa Turtles. *Ecology*. 92(11), 3010–3024.

———. (2004) *Temperature-Dependent Sex Determination in Vertabrates*. Washington: Smithsonian Books.

van Fraassen, B. C. (2008). *Scientific Representation: Paradoxes of Perspective*. Oxford: Oxford University Press.

Weisberg, M. (2006) Robustness Analysis. *Philosophy of Science*. 73, 730–742.

Wimsatt, W. (2007) *Re-Engineering Philosophy for Limited Beings: Piecewise Approximations to Reality*. Cambridge: Harvard University Press.

Wolfe-Simon, F., Switzer, B. J., Kulp, T. R., Gordon, G. W., Hoeft, S. E., Pett-Ridge, J., et al. (2011). A bacterium that can grow by using arsenic instead of phosphorus. *Science*. 332(6034), 1163–1166.

Woodward, J. (2003) *Making Things Happen: A Theory of Causal Explanation*. Oxford: Oxford University Press.

———. (2006) Some Varieties of Robustness. *Journal of Economic Methodology*. 13(2), 219–240.

5 Synthetic kinds

Kind-making in synthetic biology

Catherine Kendig and Bryan A. Bartley

Introduction

Synthetic biology may be defined broadly as the application of engineering principles to the design, construction, and analysis of biological systems.[1] The perspective that living systems can be engineered is made possible by modern biotechnologies like DNA sequencing, DNA synthesis, and genetic modification. For example, biological functions such as metabolism may now be genetically reengineered or reprogramed to produce new chemical compounds. Designing, modifying, and manufacturing new biomolecular systems and metabolic pathways draws upon analogies from engineering such as standardized parts, circuits, oscillators, and digital logic gates. These engineering techniques and computational models are then used to understand, rewire, reprogram, and reengineer biological networks and modules. But is that all there is to synthetic biology? Is this descriptive catalogue of bricolage and plug-and-play account wholly explanatory of the discipline? Do these descriptions impact scientific metaphysics?[2] If so, how might these parts descriptions inform us of what it is to be a biological kind? Attempting to answer these questions requires investigations into the nature of these biological parts as well as what role descriptions of parts play in the identification of them as the same sort of thing as another thing of the same kind.[3]

Standard biological parts with known functions are catalogued in a number of repositories (Ham et al., 2012; Madsen et al., 2016). Biological parts can then be selected from the catalogue and assembled in a variety of combinations to construct a system or pathway in a chassis microbe such as *E. coli*. Biological parts repositories serve as a common resource where synthetic biologists can go to obtain physical samples of DNA associated with descriptive data about those samples. Perhaps the best example of a biological parts repository is the iGEM Registry of Standard Biological Parts (igem.org), which grows in contributions significantly because it serves as a collaborative platform for teams competing in the International Genetically Engineered Machine Competition (iGEM). Since 2008, the Registry has grown from approximately 2,000 parts to 32,000 parts, as determined by a recent programmatic search. These parts have been classified into collections, some labeled with engineering terms (e.g., chassis, receiver), some labeled with biological terms (e.g., protein domain, binding), and some labeled with vague generality (e.g., classic, direction). Although the labeling of parts includes widely used descriptive terms, parts in the Registry are not categorized according to any formally structured terminology (Galdzicki et al., 2011).

Descriptive catalogues appear to furnish part-specific knowledge and individuation criteria that allow us to individuate them *as parts*.[4] Practitioners may ask: How do I know which part to use? Which model do I follow? Which models can I combine? and Which set of computations should I employ for my particular purpose? In order to answer this set of questions, we need to be able to track parts – or at least the names of parts – as well as their diverse multilevel descriptions that are expressed in both formal labels and natural language descriptions.

But the burgeoning array of libraries and repositories of functional parts issues in an affordance of information as well as a problem. The meaning-containing natural language descriptions and formal semantic notations that are used to annotate these parts vary across libraries and repositories (Le Novère et al., 2009). The result: biological meaning expressed in semantic annotations may not be comparable – or translatable – across different models or within different data libraries.[5]

We begin this chapter with a discussion of the nature of biological parts and their descriptions. We characterize the practice of part-making in synthetic biology by first studying how parts are characterized and tracked in different repositories. We then investigate some problems arising from the varied descriptions of parts contained in different repositories. Following this, we outline problems that arise with naming and tracking parts within and across repositories and explore how the comparison of parts across different databases might be facilitated (cf. Schulz et al., 2011). We focus on a particular set of computational models currently being sought that would allow practitioners to capture information and meta-information relevant to answering particular questions through the construction of similarity measures for different biological ontologies. We conclude by exploring the social and normative aspects of part-making and kind-making in synthetic biology.

What are parts and why does it matter?

Repositories catalogue parts. It seems straightforward enough to understand what is contained within the repository in terms of the general concept: *part*. After all, everyone knows what a *part* is. A part is something that is understood in terms of its relationship to a whole. Discovery and description of a part characterizes functional, structural, and relational attributes of that thing specified as a part. But what are we doing when we describe something as being a part? Are we simply attributing a relational property to something? If so, what does it mean to attribute the property of parthood to a constructed product, process, or mechanism within bioengineering? Usually, if we say that something is a part of something else, reference to its parthood is thought to be metaphysically grounded (e.g., its parthood is due to a particular relationship of composition, kind membership, inheritance), or indicative of our understanding of a process (e.g., which entities are involved in a pathway's functioning or evolvability). Parts may play different epistemic roles depending on how they are used and for what purpose parthood is attributed to them. For instance, part x may be named as such to identify the difference between two or more entities or events in virtue of their change over time, or to identify the role part x plays in a particular phenomenon (cf. Calcott, 2009, pp. 65–67). Calling something "part x" usually means that the thing or process being named "part x" is integral to a discrete system but also possesses a degree of individuality such that:

> [A] part is a system that is both integrated internally and isolated from its surround[ings] . . . [it] is integrated to the degree that interactions among components are many or strong, or both . . . these may be spatially distributed [such as] a hormone-mediated control system or a local population of crickets chirping in synchrony . . . Isolation is a reduction in, or termination of, integration . . . Both integration and isolation may vary continuously, and therefore the extent to which a system is a part – its degree of partness – is likewise a continuous variable.
>
> (McShea and Venit, 2001, p. 262)

The conception of parthood suggested by Calcott's and McShea and Venit's accounts seems to provide a general account of what it is to be a part and how parts might come to be known or understood. But one might complain that this does not show that the general metaphysical or epistemological concept is in any way the formalized notion of parthood used within any particular discipline within biology. However, an account of parthood similar to Calcott's and McShea and Venit's has in fact been mathematically formulated and applied to studies of modularity in native biological networks (see Guimera and Amaral, 2005). A module, in this biological context, may be thought of as a subnetwork within a larger network, a part within the whole. An ideal module has a high number of internal interactions while having few inter-modular interactions, thus integrated internally while isolated from its surroundings. When analyzed from this perspective, *E. coli*'s protein-protein interaction network is organized into a central core of modules (for cell division, DNA synthesis, and DNA maintenance), with smaller, less distinct modules arranged around the periphery (Tamames et al., 2007).

The attribution of the property of parthood also plays a significant role in synthetic biology by cataloguing known isolatable entities that can be functionally integrated into new engineered systems. Parts repositories and catalogues contain the labels and notations that effectively fix that product or process to which they refer by naming or coding. One might argue that in bioengineering, the prerequisite for something qualifying as a part is that it is named or coded. For instance, referring to some entity (e.g., a product or process) as "part x", is in some sense informative only if it is labeled as such. The part, in virtue of its label, becomes trackable and potentially comparable to other parts.

If what it means to be a part is determined by what is identified, individuated, labeled, and cached as a part, do these classificatory practices also have an impact on scientific metaphysics? We suggest that if parthood is something that is a property (or relationship) that is methodologically and epistemologically constrained by scientific practice, then these individuation and identification practices may impact how we conceive of the nature of parthood and the relationship of parts to wholes. These questions frame and motivate our discussion. We consider how parts are treated, described, used, catalogued, and compared in synthetic biology because investigating how these activities shape the conditions of individuation appears to be a necessary step – a sort of pre-classification of biological entities – that serves as a precondition for any classification that relies on a notion of part identity, on the comparison of parts as "the same part", or on "the same relationship".

From native biological kinds to synthetic biological kinds

Synthetic biology encompasses a number of overlapping approaches that include metabolic engineering and directed evolution. Both employ native biological kinds as well as synthetic biological kinds. Metabolic engineering depends on the discovery and investigation of *natural*[6] metabolic pathways, the genetic elements that control them, and on using that information to transform a suitable native pathway into a desired non-native metabolic pathway in a host organism. In addition to the reengineering of native pathways, directed evolution techniques like multiplex automated genome engineering (MAGE) (Isaacs et al., 2011) and phage-assisted continuous evolution (PACE) (Esvelt et al., 2011) accelerate evolution by introducing non-native sources of genetic variability. These directed evolutionary techniques do more than create useful laboratory variants. They expedite the evolution of organisms in ways not possible in the native process. For instance, Cedric Orelle et al. (2015) have devised a non-native orthogonal ribosome-messenger RNA system that can be used to select for mutations that would be lethal in the native process. These synthetic ribosomes can potentially be used for production of non-native protein analogs in *E. coli* without disrupting production of native proteins necessary for cell growth and viability.

Not only do these new technologies provide opportunities for the reengineering of biological systems in synthetic biological practice, but also they usher in new articulations of parthood that may have wider metaphysical and epistemic impact. For example, reengineering synthetic metabolic machines introduces more than just ribosomes with new functions, it introduces new categories of what it means to be a ribosome. Rather than a ribosome being the kind of thing that we can categorize as either one whose production of proteins is either native or non-native, Orelle et al. (2015) showed that they may also be categorized as producing both native and non-native proteins. That is, the sorts of ribosomes or categories of ribosomes change depending on how they produce proteins. Metabolic engineering and genome editing techniques like MAGE, PACE, and CRISPR/Cas9 may suggest a reconfiguration of what it is to be a natural kind – or perhaps what might be better referred to as a "synthetic kind". But what does it mean to call something a "synthetic kind"? We might conceive of a *synthetic kind* as a form of life (or at least a life-like thing) whose construction is the result of human-assisted engineering. In this way, we are partitioning those entities which might be candidates to be described as *synthetic biological kinds*, in virtue of their origin. For instance, we might suggest that an *E. coli* population harboring a synthetic DNA plasmid might be considered to be a synthetic kind.[7] Synthetic DNA may be derived from entirely non-living sources, but may be considered living once integrated into the host. This system consists of a combination of the wild-type host organism and the synthetic plasmid. Because of the different causal origins of the system – as being the result of both engineered and evolved parts and processes evolved – it owes its existence to both engineered and native, wild-type sources. We consider there to be several broad categories of synthetic biology, the consideration of whose diverse causal origins may suggest a new synthetic (or semi-synthetic) category of biological kindhood:

1 Directed evolution and artificial selection of new organisms.
2 Inserting non-native DNA (biological parts) into a host, or genome editing, for example, via CRISPR/Cas9.
3 Minimal synthetic life. No pre-living ingredients are used to create minimal synthetic life.

A *synthetic biological kind* does not necessarily imply that the kind being specified as such qualifies as an *unnatural kind*. Synthetic kinds may use natural biological kinds (or what might be better termed in this context "native biological kinds"), as building blocks. For example, an organism widely considered to be a synthetic organism may have either a considerable amount of its genome reengineered or none at all. In many cases, a native organism is altered by introducing DNA molecules called plasmids which replicate independently of the host organism's genome. The synthetic plasmid may consist of multiple genes native to other organisms that are recombined to create novel, synthetic phenotypes. The host organism is often referred to as a "chassis", which in engineering refers to a frame upon which other components are assembled. Such synthetic organisms are the combination of both synthetic and native kinds. A synthetic biological kind may, therefore, be a combination of different native biological kinds, or a combination of a native biological kind with a synthetic kind.

   Table 5.1 shows some examples of putative native kinds in genetics and their corresponding putative synthetic kinds in synthetic biology.

Table 5.1 Putative native and synthetic kinds

Native kind	Synthetic kind
Fluorescent protein	Reporter protein
Ribozyme	Insulator
Repressible promoter	Inverter

Focusing on how biological entities are described as parts, recorded, and labeled as such provides the necessary conditions under which an entity can be identified and individuated. The possibility of comparing parts once trackable by their label or description affords the possibility of grouping "the same" parts into types or kinds. Within synthetic biology, these types or kinds do not fit perfectly within either what are generally referred to as "technological kinds", nor do they fit with what are generally understood to be natural kinds (cf. Schyfter, 2012). If they are neither fully technological nor natural biological kinds, does this pose a problem for conceiving of these entities as kinds at all, or is there a third option? To argue that they fail to have the requisite properties of either technological or natural biological kinds, is there something that both have in common that is missing from synthetic kinds, for example, some sort of kind-maker that by virtue of being synthetic, they lack, but technological and natural biological kinds possess? If this is the case, what is it? Determining what kinds (i.e., technological kinds or natural biological kinds or something else) exist in synthetic biology may ultimately rest on the discipline itself and how it is understood in relation to other biological and engineering disciplines.

The discipline of synthetic biology is sometimes conceived of by practitioners and detractors as a subset of functional biology and as such is characterized as an application-based, or technology-based mode of understanding that seeks to explain how something works. It has also been characterized as evolutionary biology due to its attempt (especially in protocell creation) to answer why-questions: seeking why (rather than how) biological pathways, devices, and parts work. This difference in the attribution of goals, products, and techniques depending on what types of questions are being asked make the categorization of synthetic biology as a disunified discipline unsurprising (cf. Keller, 2009; Bensaude Vincent, 2013). Its growing epistemic and methodological toolkit seems likely to continue apace – the result of sourcing and modifying techniques from biology, chemistry, computer science, mathematics, and engineering.[8]

But relying on this dichotomy of functional and evolutionary biology, of how- and why-questions does not seem entirely justified – or at least is not always elucidatory – within synthetic biology. Knowledge-seeking questions within synthetic biology do not focus purely on how-questions directed for the purpose of modifying function. One of us suggests that such dichotomizations fail to identify the union of how- and why-questions, their mode of investigation, and categorization and kind-making – or what Kendig (2016) calls "kinding" – practices typified by synthetic biology. The making or constructing of material objects, mechanisms, processes, or pathways; the theoretical construction of models and algorithms; as well as the devising of repeatable methods and techniques being made in synthetic biology are all instances of kinding – where kinding is understood as the epistemological and ontological activities within the practice of synthetic biology and by which the categories of that discipline or subdiscipline are configured. The outputs of these practices – the products of diverse synthetic biological research aims – are exchangeable and repeatable activities that represent, explain, and advance our understanding of the relation of parts and wholes, the manipulation of developmental pathways, and the nature of biological function-ing and organization. This is exemplified in the engineered construction of func-tional parts, processes, pathways, devices, and systems (Brent, 2004; Endy, 2005). Understanding of these functional systems is borne out in their decomposition, manipulation, co-option, and construction. This philosophy of synthetic biology is often summarized through the words of the physicist Richard Feynman: "What I cannot create, I do not understand" – words that he left scrawled on a chalkboard at the time of his death. The type of synthetic biology focused on the engineered construction of functional parts, processes, pathways, devices, and systems is in the business of producing standardized parts, devices, pathways, and modules with known functions. In their approach to kinding, synthetic biologists attribute purpose, intent, and design to biological parts, while traditional biologists often avoid describing natural kinds in such teleological terms (see Table 5.1).

The practice of cataloguing and describing standard biological parts with known functions for future reference by practitioners appears to be, if not the actual making of these parts as parts, at least the conditions under which they may be used as parts. It would then seem uncontroversial to suggest that the imperfect fit of the parts and processes of synthetic biology into technological or biological kinds is because that which is kinded is epistemologically heterogeneous. The explanations and descriptions contained within parts repositories suggest that it may be more profitable to conceive of synthetic biology as a discipline that is epistemologically, ontologically, and methodologically hybrid. Synthetic biology takes on more than engineering practices and methodology. It brings an engi-neering problem-solving attitude to bear on biological problems that makes the production of synthetic biological products possible. But in addition to this hybrid engineering-biology epistemology comes a new hybrid ontology. This ontology includes both new entities – new kinds of native-non-native sorts of being, but also new practices of sorting and kinding the products of synthetic biology. In this way, synthetic biological research, as well as the synthetic products and pro-cesses that it creates, sits in a liminal position between engineered "technologi-cal kinds" and so-called natural (or "native") biological kinds. If this liminality exists, it presents difficulties with regard to the notion of kindhood if kindhood is understood (as it classically is), to be a property of the contents of the world as already there for us. Synthetic biological research and the non-native products and processes that arise from it would seem to suggest a notion of kindhood that does not rely on the pre-carved up contents of the world. Instead, it appears to involve the role of the carvers (e.g., the synthetic biology practitioners as kind-makers or kind-locators), as well as the ascribers and cataloguers of the joints partitioning the wholes.

Naming and tracking parts within and across repositories

Naming and tracking is pervasive in all fields of biology but seems to play a particular role in the history and philosophy of taxonomy, comparative anatomy, physiology, genomics, and synthetic biology where discovering something is the

same part is crucial to resolving identity, isomorphism, or phylogenetic conti-
nuity. To investigate how entities are discovered to be the same, how they are
tracked, and why it matters seems to require ineliminable reference to the criteria
of identity being used to assess whether x is the same sort of thing as y. The individuation of parts and the notion of parthood relies on an unavoidably context-
dependent notion:

> [W]e can individuate the different theoretical perspectives, T_i, applicable to a
> system. Each of these T_i perspectives implies or suggests criteria for identifi-
> cation and individuation of parts, and thus generates a decomposition of the
> system into parts. These decompositions, $K(T)_i$, I will call K-decompositions.
> The different K-decompositions may or may not give spatially coincident
> boundaries for some or for all of the parts of the system.
>
> (Wimsatt, 2007, p. 182)

The criteria for individuation must always be specified within a particular theoretical framework (T_i) that contains within it the criteria to be used for the identification of parts qua parts for that particular K-decomposition, $K(T)_i$. But if we
grant that parthood is a property that is ineliminably context dependent, what
does this mean if we want to track a part, qua part, across different theoretical systems? Is this even possible? In order to attempt to answer this question, we look
to the development and use of networked information systems and what is often
described as "ontology engineering" (Gruber, 2009).

The practice of ontology engineering certainly borrows some inspiration from
philosophical ontology, though there are some important distinctions (Zúñiga,
2001). In the context of information systems, ontologies are specified in terms
of both natural language descriptions and machine-readable code. For example,
the Sequence Ontology (Eilbeck et al., 2005), which describes regions of DNA
sequences, defines a gene as "a region (or regions) that includes all of the sequence
elements necessary to encode a functional transcript". In terms of database infra-
structure, this natural language description is less useful, and the gene concept
might be internally referenced by its unique accession number, SO:0000704. At
the level of software and database engineering, computational ontologies help
machines ask queries, interpret answers, and integrate data from multiple sources.
Today, many large-scale, collaborative enterprises, including engineering, business operations management, and life sciences, are currently supported by a cloud
of networked databases and software tools which communicate through ontologies (Rachuri et al., 2008; Wilkinson et al., 2016).

Ontologies identify, describe, and classify the material objects, entities, or fundamental concepts encountered in a field of practice. Ontologies very often classify
objects in terms of what we might call their parthood (e.g., compositional properties) and kindhood (e.g., functional class membership, structural types, taxonomy).
For example, the Sequence Ontology specifies that a promoter, coding_sequence,
and UTR are each part_of a gene, while transcription_regulatory_region,
translation_regulatory_region, and replication_regulatory_region are each a

kind_of regulatory_region. At a high level, ontologies are catalogues of human knowledge and experience, while at a low level, they represent this knowledge in a data format that can be understood by software applications, databases, and information systems. Ontologies have become critical to manage the surfeit of information in many *Big-Data* approaches in the life sciences.

Many such classification systems in the life sciences can be explored interactively on the web. An important resource for biomedical ontologies, the BioPortal, currently hosts 597 biomedical ontologies (bioportal.bioontology.org). The first biomedical ontologies were developed by the Open Biomedical Ontologies consortium and hence were expressed in a data format called OBO (Ashburner et al., 2000). But the OBO format suffers from ambiguities, making it difficult for users to share data and integrate knowledge (Golbreich et al., 2007). Therefore, many research communities have begun adopting the Web Ontology Language (OWL), a formal description logic, as the preferred machine-readable format for ontologies. This example illustrates the practical difficulties involved when integrating theoretical frameworks.[9] However, our focus in this chapter is the technical aspects of developing and using ontologies. That is, focusing on how information from different parts repositories is shared across research groups when the repositories classify parts in different ways using different language to capture different concepts.

Standardizing and translating annotations of different ontologies

Given the multiplicity of repositories utilizing different ontologies, the problem that arises is that the labeling of parts of genetic data are known within, but not across, ontologies. Tracking the same synthetic part in one descriptive ontology relies on knowing how to translate one annotation within one ontology to that of another. Despite standardization being thought to be the cornerstone of synthetic biology, it has remained elusive with regard to obtaining a standardization among the descriptions of various synthetic parts. Models are published continuously; lists of parts are updated and corrected. However, descriptions are user-driven and may be functionally specific (Richards, 2006). They may focus on one level of organization in one list and another in a separate database: "Annotations may . . . relate cell compartments to Gene Ontology entries and small chemical compounds to entries from ChEBI" (Schulz et al., 2011). A system for tracking meaning of descriptive labels across ontologies and fixing the reference class of the parts to which the descriptive labels apply is actively being sought.

While the iGEM Registry of Standard Biological Parts continues to be the largest biological parts repository, other synthetic biology repositories are being developed that can share data and support queries with each other, forming a "web of registries" (Ham et al., 2012; Madsen et al., 2016). In order for software tools to communicate with these knowledge-bases, a prior, formal agreement is required that specifies how knowledge will be transmitted. The Synthetic Biology Open Language (SBOL) is a data exchange standard which was developed to support the vision of a web of registries for synthetic biology (Roehner et al., 2016; Galdzicki et al., 2011). In addition to being a machine-readable language, SBOL also models concepts from synthetic biology in use across different disciplines. The data model is the result of community efforts to integrate knowledge at the level of engineering and biology. Biological parts are represented using classes such as Components, Sequences, Modules, and Models, and these classes are further described in terms of their properties and their relationships with each other. This data model facilitates reasoning about the modular composition of biological structure and function, abstraction hierarchies, and automated workflows, which are foundational principles of synthetic biology (Endy, 2005). But SBOL also represents biological knowledge compiled from diverse sources. Therefore, SBOL developers must consider how SBOL interfaces with biological knowledge-bases, which may communicate through standards like GenBank (www.ncbi.nlm.nih.gov/genbank/) (Zundel et al., 2017) or the Research Collaboratory for Structural Bioinformatics Protein Data Bank (RCSB PDB) (www.rcsb.org/pdb/home/home.do).

Within SBOL, biological kinds are coded or annotated with different descriptive ontologies, such as the Sequence Ontology (Eilbeck et al., 2005). A synthetic biologist may use terms from this ontology to specify which kind of parts should be used for a particular design. The Sequence Ontology describes both natural and engineered genetic parts and kinds. The category structure of the Ontology is built around non-mutually exclusive category tags, which allows users to classify a feature of a part as experimental, biological, biomaterial, or some combination thereof. Because of this, Sequence Ontology has a way of making distinctions between natural and engineered parts by relying on the descriptive category. However, these categories (and the descriptions contained therein) are not arranged in a purely hierarchical way. That means that something could be classified as an engineered and as a natural thing as these categorical descriptions may overlap with one another, rather than in a relationship of containment or exclusion. Because of the different use of third-party ontologies and different ways SBOL may interface with different knowledge-bases, SBOL developers must consider how both native biological parts and synthetic biological parts are classified in an attempt to integrate diverse fields of knowledge. However, it remains an open challenge to integrate and align the semantics of SBOL, iGEM Registry, and the Sequence Ontology toward a more coherent and integrated understanding of synthetic biology. Some might ask whether there is indeed evidence that integration is a goal at all.[10] There are at least two ways to respond to this concern – a pragmatic (or technical) one and a theoretical one. The first is a quite pragmatic and emphatic: yes. There is evidence that integrated understanding in synthetic biology is a goal because in order to pursue and resolve research problems and development solutions within science requires large groups of scientists working together, for example, experts in mathematical modeling and experimental testing. The second is a theoretical or normative question of whether integration should be a goal.[11]

Comparing biological parts

The various ontologies and repositories (e.g., iGEM Registry, SBOL, SO, RCSB PDB) appear to act as ontologies at least in terms of their functional role as catalogues defining parts and serving as resources for the discipline of synthetic biology and biological engineering. In doing so, they define a systematic classification of parts on the basis of their functioning which may then be used to determine if one part that a practitioner discovers is the same kind of part as another already labeled and described in a repository. But how might two parts be compared in order to determine whether or not they are the same part? This is possible through quantitative metrics that measure how closely related parts are to each other. These are called "semantic similarity measures" (Pesquita et al., 2009).

One of the most recent approaches to standardize and translate annotations of different ontologies is a tree-based system called "phylosemantics" (Bartley et al., 2017). The term "phylosemantics" is a portmanteau of phylogenetics and semantics. Phylogenetics is a well-established field that infers evolutionary relationships between organisms on the basis of similarities in their DNA sequences. Phylogenetics draws on an even older idea from information theory and computer science, that of an edit distance, or Levenshtein distance which measures how similar two strings of data are to each other (Bartley et al., 2017). This measure defines the lexical distance between two strings by counting the number of insertions, deletions, or substitutions necessary to transform one string into another.[12] Relying on the metaphor of the genetic code being "written" in four letters (AGTC) representing the four amino acids – Adenine, Guanine, Thymine, and Cytosine – the evolutionary distance between two DNA sequences is inferred by assuming a probability that one letter of the genetic code will transform or mutate across a transformational series from one into another.

"Semantic" refers to the meaning of natural language and asserts that meaning can be not only studied but also quantified. Though speaking of semantics as a quantitative science might strike the reader as controversial at first, a familiar example demonstrates the effectiveness of quantifying meaning in the sciences. Take for instance, the Wong-Baker Faces pain assessment scale commonly used in doctors' offices (Wong and Baker, 1988). The series of six faces begins with a drawing of a face with a broad smile and upturned eyebrows. Indicated below this drawing is "0" and underneath, the words "No Hurt". To the right of this face is a second that is also smiling – but less so than the first and with eyebrows that are not raised as much. Under the second image is "2" and the words "Hurts Little Bit". The scale progresses with a face depicting "Hurts Little More" – a mouth that is drawn as a straight line and ascribed "4". Face 6, "Hurts Even More" is a sad face with downturned eyes and eyebrows that turn down on the sides. 8, "Hurts Whole Lot" has even more downward turned mouth, eyes, eyebrows that indent the face as if on the verge of tears. The series ends with "10", "Hurts Worst", represented with a mouth in an upside-down U and tears that are streaming down from the drawing's eyes.

Quantitative assessment and the employment of computers to analytically model and explore the semantics of language is not new, but it is an approach that has led to the generation of a number of new computational fields of inquiry such as natural language processing, artificial intelligence, and the ontology engineering discussed earlier (Léon, 2007). Computational ontologies have been explored as a means to structure knowledge and organize data, especially on networked machines. Ontologies have also been developed for diverse applications in the life sciences (Smith et al., 2007). The reference to semantics in the name "phylosemantics" is intended to indicate that these kinds of semantic analyses might yield new and interesting insights for synthetic biology as well.

Akin to its partial namesake phylogenetics, phylosemantics also relies on trees. Phylosemantic trees are obtained by comparing real-language descriptions of DNA sequences, and calculating a semantic distance between the meaning of these real-language descriptions. During the process of studying DNA sequences, synthetic biology practitioners often use software tools to add descriptive annotations to their data. These annotations encode meaning, which can be formally classified using ontologies. Semantic similarity measurements have been used for diverse analyses in the life sciences. It has become a valuable tool for validating results drawn from biomedical studies such as gene clustering, gene expression data analysis, prediction and validation of molecular interactions, and disease gene prioritization (Pesquita et al., 2009). Semantic similarity can also be used to classify mechanistic models of biochemical systems by employing a tree diagram to classify different theoretical models of a MAPK pathway in cancer signaling (Schulz et al., 2011). The diagram reveals the knowledge space of hypotheses that are intended to explain the signaling mechanism.

The phylosemantic approach continues this trend to study the relationship between genetic architecture and phenotypic behavior. It allows practitioners to classify genetic structure, or architecture, based on descriptive abstractions and can be used to classify measurements, identify patterns, and discover design-rules for genetic architectures.[13] However, the approach also aims to respond to the open challenge to integrate and align the semantics of SBOL, iGEM parts categories, and the Sequence Ontology toward a more integrated understanding of synthetic biology. Phylosemantics compares different descriptive labels used to annotate genetic data and classify synthetic biological parts (e.g., combinatorial promoters) to find out how similar they are to one another. The tree-like model that results from the comparison reveals the semantic distance of different engineered genetic architectures – a technique that is based on *semantic similarity* (see Pesquita et al., 2009). Relying on quantitative similarity measures between different descriptions allows practitioners to compare labels or annotations in terms of different biological[14] (e.g., seeking to investigate how a highly conserved metabolic process is retained over evolutionary time in a variety of species) or mathematical[15] (e.g., the use of trees to compare network structures) measures depending on which suit the needs of the investigator and her or his system of focus. Construction of a tree allows practitioners to rank and classify and identify semantic overlap or complementarity between parts or concepts used in different databases. One attempt to standardize data is Dublin Core Metadata. It provides a supra-level ontology of descriptions of relations (e.g., "isVersionOf") and parts (e.g., "hasPart") that allow practitioners to label their content. It also includes a unique "identifier" for the content – a formal name, so that it can be named and tracked (dublincore.org).

Social and normative aspects of part-making and kind-making in synthetic biology

The explicitly practice-based route of investigation that we have pursued in this chapter has so far focused on scientists' activities of acquiring, describing, cataloguing, and semantically aligning knowledge of parts within synthetic biology. This approach presents an alternative way of exploring traditional metaphysical questions in philosophy such as What are the kinds of things synthetic biology produces? and What is the relationship between parts and wholes? It also introduces new questions about what it is to be a part or indeed the property of partness itself within a new field; questions not just of interest to metaphysicians of science. Instead of seeking to understand the use of epistemological and ontological categories in practice from the premise that their existence can be known *a priori* or contained within the theoretical framework of the discipline that uses them without reflection on the use made of them by practitioners, this approach runs in the opposite direction. It instead suggests that these categories of existence come into being in practice and from these categories-in-use *cum* theoretical concepts, which then reciprocally feed back into the use to which they are made in practice.

Identifying the work of scientists as being not just the subject of epistemology but also of an engagement with the valuation of knowledge is not new. Marjorie Grene (1966, p. 223) emphasized that the work of scientists was "an instance of the recognition of responsible persons, a performance of the same general kind as the recognition of patterns, individuals, or persons". Identifying the centrality of this social aspect of scientific work and the conditions under which knowledge-producing labor takes place, she refers to the work of scientists as occurring within "social enterprises" (Grene, 1985). More recently, these social and normative aspects have been described as the cumulative knowledge of groups as scientific "social cognition" (Longino, 1990); referred to in virtue ethics as subject-centered or "agent-based" interactions (Swanton, 2003); discussed in terms of the discipline-specific knowledge generated within "systems of practice" (Chang, 2012); and cashed out as the "repertoires" that promote social engagement, communication, data management, and organization of resources that make collaboration possible (Leonelli and Ankeny, 2016). We build on this formidable work, focusing on the social aspect of scientific investigation and the notion of a system of practice in order to identify the use of concepts by practitioners and explore how their utilization of these in practice suggests an ongoing and reciprocal tuning of concept to practice and practice to concept within synthetic biology. Doing so, we briefly discuss how normative valuations effectively shape the sorts of things (e.g., parts, kinds, processes) thought to be the subject of study for the discipline of synthetic biology. Our discussion focuses specifically on the normative valuation of practitioners utilizing SBOL. We begin by introducing the epistemic principles that have been used within computational ontologies.

Thomas Gruber, widely considered to be the founder of modern computational ontologies, identified the following epistemic principles as being central to the development of ontologies (Gruber, 1995): clarity, coherence, extendibility, minimal ontological commitment, and minimal encoding bias. Evidence for how these epistemic ideals may be instantiated can be found by looking more closely at the SBOL data model we introduced earlier. The SBOL data model attempts to exhibit clarity by emphasizing key engineering concepts like Models, Components, and Modules. The specification values coherence, evidenced by rules which enforce logical consistency of the data. However, not all of the classes in the data model describe concepts from synthetic biology, and some of the class names reflect concepts related to their data encoding. For instance, one SBOL class is called a ComponentDefinition, which mixes ideas at the implementation level with ideas at the knowledge level. The result is that Definitions and Instances are concepts related to how data is stored in databases, but which do not reflect knowledge at the domain level of synthetic biology, an example of encoding bias.

But these kinds of computational ontologies are useful for more than just helping machines speak to one another. Usually the end-user is a human who must interpret and make decisions based on data. How the data are structured affects how they are presented to the user and how the user interacts with the data. For this reason, many suggest that ontologies should be designed so that software tools present simple and intuitive interfaces to the user (Mi and Thomas, 2011; Leonelli, 2012; Leonelli and Ankeny, 2016). For instance, communities that develop ontologies and standards might benefit from considering which epistemic principles they most value as they work toward a consensus model of their knowledge domain.

As a growing field of knowledge that is likely to generate new parts and new descriptions, constant reevaluation of a poorly designed classification scheme appears necessary. It is in this process that epistemic considerations have played an important role. In attempting to answer the question of how we know when an ontology is an accurate representation of a knowledge space, a well-suited ontology may be able to map where a field has been and anticipates where it might head. Characterization of the knowledge space within synthetic biology often comes from borrowing the methods and principles of other disciplines, such as engineering. Engineering principles distinguish the field of synthetic biology apart from its historical antecedents in genetics, molecular biology, and biophysics. These principles include standardization, modularity, and abstraction. If scientific practice may in any sense inform metaphysics, we might do well to consider how the principles of synthetic biology (e.g., abstraction, modularity, and standardization) impact activities of kind-making and part-making. These forms of scientific metaphysics may not be solely restricted to the metaphysical categories of synthetic biology. Perhaps they may apply even more generally to other disciplines focusing on synthetic kinds (see Table 5.2).

Table 5.2 Synthetic biology principles aligned to categories of knowledge-making

Principle of synthetic biology	*Epistemological category of knowledge-making*
Abstraction	kind-making
Modularity	part-making
Standardization	part-making and kind-making

Taking this suggestion seriously, what does this mean for the work of synthetic biology practitioners, or scientists more generally? And more pointedly, how are these principles used within the work of scientists? Scientists rely on socially accepted protocols as sets of explicit and implicit rules of action. However, the action of individual agents within the social group of scientists is not the work of idealized decision-makers, but seems to be biased on individuals whose beliefs are not entirely self-generated. Grene's (1966) view seems underpinned by exactly this sort of agent-based view. In so doing, it investigates a perspective where action is always from a standpoint, but that the standpoint is located in social networks of knowledge-producing practices. It is a place from which the agent acts that appears to be itself a precondition for personal action.

The social nature of scientific practice that Longino suggests also emphasizes the intimate connection between the character of inquiry and the goals of individual scientists. As such, it is also prerequisite for what she refers to as "social cognition" – the idea that scientific inquiry is not simply an individual pursuit but an epistemic activity relying on the intersubjectivity of critical dialogue within scientific study (Longino, 1990). In doing so, she follows Grene (1966, 1985) in fleshing out the nature of what it means to be a social enterprise and what Chang (2004, 2012) later identifies as a system of practice. According to Chang, the search for an agent-free or context-free set of categories is simply one that is ill-founded in philosophy:

> A system of practice is formed by a coherent set of epistemic activities performed with a view to achieve certain aims . . . [A]s with coherence of each activity, it is the overall aims of a system of practice that define what it means for the system to be coherent. The coherence of a system goes beyond mere consistency between the propositions involved in its activities: rather, coherence consists in various activities coming together in an effective way toward the achievement of the aims of the system.
>
> (Chang, 2012, p. 16)

The development of parts repositories, the meaning of what it is to be a part in synthetic biology, and the modification by the discoveries, methods of cataloguing, tracking, and comparing parts in use to both define and redefine parts has also redefined the conception of parthood employed within synthetic biology. As such, it is one example of a "coherent set of epistemic activities performed to achieve a certain aim" that has also defined the concepts in use through the practice of their implementation. We have described in this chapter how this process has been manifest in the use made of repositories by synthetic biology practitioners to catalogue their discoveries and in their employment of it for future biological engineering of systems using the parts contained within them.

We have suggested that the use of repositories to catalogue the discoveries and descriptions of as-yet unclassified things provides an epistemic framework necessary for the pre-classification of parts. Their parthood is attributed once described or used successfully within a synthetic system by practitioners. We have explored the importance of knowing that something is the same part as something else in two distinct but interconnected ways: the identification of parts in terms of representing knowledge and catalogued within repositories, and the merging of fields.[16] It is for this reason that our explicit aim in this chapter has not been to discuss completed ontologies – once made, discovered, or identified – but instead to focus on the preliminary steps and processes of knowledge production which are prerequisite to the construction or identification of ontologies of parts within synthetic biology.

Acknowledgments

This material is based in part upon research supported by the National Science Foundation under Cooperative Agreement No. DBI-0939454 and the BEACON Center for the Study of Evolution in Action. Catherine Kendig and Bryan A. Bartley are members of the BEACON Center for the Study of Evolution in Action. Development of SBOL is supported by NSF DBI-1355909.

Notes

1 See especially (Bartley et al., 2017) and (Kendig and Eckdahl, 2017).
2 Traditional metaphysics investigates the general structure of reality and the basic units of existence in the universe (e.g., entities, processes, kinds) from a largely unempirical vantage point. *Scientific* metaphysics focuses on how scientific practices, concepts, tools, classifications, modes of experimentation, testing, and protocols used within various scientific disciplines shapes, and in some cases generates, these basic units (see Kendig, 2016).
3 In later sections, we will elaborate on the identification of parts as the same sort of part within synthetic biology by employing the notion of *semantic similarity*.
4 We follow the convention within synthetic biology and refer to these as *parts*. But, to be sure, their parthood within the repository is best understood as *potential parthood*. Prior to being used within a natural or engineered system, it is not part of anything, yet.
5 For example, BioModels Database www.ebi.ac.uk/biomodels/, PathGuide http://pathguide.org.
6 A *natural* metabolic pathway is typically referred to in the synthetic biology literature as a *native* metabolic pathway, so that it is that which is *native* that is the contrast class of that which is *synthetic*.
7 *E. coli* is just one example among several possible microbial chassis organisms for synthetic biology, as there are diverse laboratory models under investigation.
8 See also Sprinzak and Elowitz (2005) and Morange (2009) for further discussions of discipline-building in synthetic biology.
9 While some practitioners suggest that ontologies are theoretical frameworks, others argue that they are not. For some, ontologies are simply caches of part-specifying information that can be deposited and retrieved and are the sorts of things that do not require any additional metaphysical claim that they represent or constitute a theoretical framework (Guarino, 1995).
10 We thank the editor, Julia R. S. Bursten, for pointing out this possible concern.
11 We do not argue for a theoretical integration in this chapter.
12 Brett Calcott explains this transformation nicely by the use of a word puzzle in which the word "scale" is transformed one letter at a time into the word "plume", with the

caveat that each word in the transformational series is also a word: "scale – scalp – scamp – stamp – stump – slump – plump – plume" (Calcott, 2009, p. 53).
13 Bartley et al. (2017) demonstrated how ontologies can be used to classify and compare synthetic DNA sequences.
14 See (Pinter et al., 2005).
15 See (Pinter et al., 2005).
16 Our approach is intended to be a friendly extension to (or at least align with) Wimsatt's (2007) discussion of parts and the comparison across different K-compositions.

References

Ashburner, M., Ball, C. A., Blake, J. A., Botstein, D., Butler, H., Cherry, J. M., Davis, A. P., Dolinski, K., Dwight, S. S., Eppig, J. T. & Harris, M. A. (2000) Gene Ontology: Tool for the Unification of Biology. *Nature Genetics.* 25(1), 25.
Bartley, B. A., Galdzicki, M., Cox, R. S. & Sauro, H. M. (2017) Mapping Genetic Design Space with Phylosemantics. Proceedings of the 9th International Workshop on Bio-Design Automation, University of Pittsburgh. Retrieved from www.iwbdaconf.org/2017/docs/IWBDA_2017_Proceedings.pdf.
Bartley, B. A., Kim, K., Medley, J. K. & Sauro, H. M. (2017) Synthetic Biology: Engineering Living Systems from Biophysical Principles. *Biophysical Journal.* 112(6), 1050–1058.
Bensaude Vincent, B. (2013) Discipline-Building in Synthetic Biology. *Studies in the History and Philosophy of the Biological and Biomedical Sciences.* 44(2), 122–129.
Brent, R. (2004) A Partnership between Biology and Engineering. *Nature Biotechnology.* 22, 1211–1214.
Calcott, B. (2009) Lineage Explanations: Explaining How Biological Mechanisms Change. *British Journal for the Philosophy of Science.* 60, 51–78.
Chang, H. (2004) *Inventing Temperature: Measurement and Scientific Progress.* New York: Oxford University Press.
Chang, H. (2012) *Is water H2O?* New York: Springer.
Eilbeck, K., Lewis, S. E., Mungall, C. J., Yandell, M., Stein, L., Durbin, R. & Ashburner, M. (2005) The Sequence Ontology: A Tool for the Unification of Genome Annotations. *Genome Biology.* 6(5), R44.
Endy, D. (2005) Foundations for Engineering Biology. *Nature.* 438(7067), 449.
Esvelt, K., Carlson, J. & Liu, D. (2011) A System for the Continuous Directed Evolution of Biomolecules. *Nature.* 472, 499–503.
Galdzicki, M., Rodriguez, C., Chandran, D., Sauro, H. M. & Gennari, J. H. (2011) Standard Biological Parts Knowledgebase. *PLoS One.* 6(2), e17005. https://doi.org/10.1371/journal.pone.0017005.
Golbreich, C., Horridge, M., Horrocks, I., Motik, B. & Shearer, R. (2007) OBO and OWL: Leveraging Semantic Web Technologies for the Life Sciences. *The Semantic Web.* 169–182.
Grene, M. (1966) *The Knower and the Known.* Berkeley: University of California Press.
Grene, M. (1985) Perception, Interpretation, and the Sciences: Toward a New Philosophy of Science. In D. J. Depew & B. H. Weber (Eds.), *Evolution at a Crossroads* (pp. 1–20). Cambridge, MA: MIT Press.
Gruber, T. (1995) Toward Principles for the Design of Ontologies Used for Knowledge Sharing? *International Journal of Human-Computer Studies.* 43(5–6), 907–928.
Gruber, T. (2009) *Ontology.* Encyclopedia of Database Systems. Berlin/Heidelberg: Springer-Verlag.

Guarino, N. (1995) Formal Ontology, Conceptual Analysis and Knowledge Representation. *International Journal of Human-Computer Studies.* 43(5–6), 625–640.

Guimera, R. & Amaral, L. A. (2005) Functional Cartography of Complex Metabolic Networks. *Nature.* 433(7028), 895.

Ham, T. S., Dmytriv, Z., Plahar, H., Chen, J., Hillson, N. J. & Keasling, J. D. (2012) Design, Implementation and Practice of JBEI-ICE: An Open Source Biological Part Registry Platform and Tools. *Nucleic Acids Research.* 40(18), e141.

Isaacs, F., Carr, P., Wang, H., Lajoie, M., Sterling, B., Kraal, L., Tolonen, A., Gianoulis, T., Goodman, C., Reppas, N., Emig, C., Bang, D., Hwang, S., Jewett, M., Jacobson, J. & Church, G. (2011) Precise Manipulation of Chromosomes in Vivo Enables Genome-Wide Codon Replacement. *Science.* 333(6040), 348–353.

Le Novère, N., Hucka, M., Mi, H., Moodie, S., Schreiber, F., Sorokin, A., Demir, E., Wegner, K., Aladjem, M. I., Wimalaratne, S. M. & Bergman, F. T. (2009) The Systems Biology Graphical Notation. *Nature Biotechnology.* 27(8), 735–741.

Léon, J. (2007) From Universal Languages to Intermediary Languages in Machine Translation. In *History of Linguistics 2002: Selected Papers from the Ninth International Conference on the History of the Language Sciences, 27–30 August 2002, São Paulo-Campinas.* Vol. 110 (p. 123). Amsterdam: John Benjamins Publishing.

Leonelli, S. (2012) Classificatory Theory in Data-Intensive Science: The Case of Open Biomedical Ontologies. *International Studies in the Philosophy of Science.* 26(1), 47–65.

Leonelli, S. & Ankeny, R. (2016) Repertoires: A Post-Kuhnian Perspective on Scientific Change and Collaborative Research. *Studies in History and Philosophy of Science Part A.* 60, 18–28.

Longino, H. (1990) *Science as Social Knowledge: Values and Objectivity in Scientific Inquiry.* Princeton: Princeton University Press.

Keller, E. F. (2009) Knowledge as Making, Making as Knowing: The Many Lives of Synthetic Biology. *Biological Theory.* 4(4), 333–339.

Kendig, C. (2016) Activities of Kinding in Scientific Practice. In C. Kendig (Ed.), *Natural Kinds and Classification in Scientific Practice.* Abingdon and New York: Routledge.

Kendig, C. & Eckdahl, T. T. (2017) Reengineering Metaphysics: Modularity, Parthood, and Evolvability in Metabolic Engineering. In A. M. Ferner & T. Pradeu (Eds.), Special Issue: Ontologies of Living Beings: Philosophy, Theory, and Practice in Biology. *Philosophy, Theory, and Practice in Biology*, Vol. 9 (p. 8).

Madsen, C., McLaughlin, J. A., Misirli, G., Pocock, M., Flanagan, K., Hallinan, J. & Wipat, A. (2016) The SBOL Stack: A Platform for Storing, Publishing, and Sharing Synthetic Biology Designs. *ACS Synthetic Biology.* 5(6), 487–497.

McShea, D. & Venit, E. (2001) What Is a Part? In G. Wagner (Ed.), *The Character Concept in Evolutionary Biology* (p. 259). San Diego: Academic Press.

Mi, H. & Thomas, P. D. (2011) Ontologies and Standards in Bioscience Research: For Machine or for Human. *Frontiers in Physiology.* 2, 5.

Morange, M. 2009. Synthetic Biology: A Bridge between Functional and Evolutionary Biology. *Biological Theory.* 4(4), 368–377.

Orelle, C., Carlson, E., Szal, T., Florin, T., Jewett, M. & Mankin, A. (2015) Protein Synthesis by Ribosomes with Tethered Subunits. *Nature.* 524, 119–124.

Pesquita, C., Faria, D., Falcão, A. O., Lord, P. & Couto, F. M. (2009) Semantic Similarity in Biomedical Ontologies. *PLOS Computational Biology.* 5(7), e1000443.

Pinter, R., Rokhlenko, O., Yeger-Lotem, E. & Ziv-Ukelson, M. (2005) Alignment of Metabolic Pathways. *Bioinformatics.* 21, 3401.

Rachuri, S., Subrahmanian, E., Bouras, A., Fenves, S., Foufou, S. & Sriram, R. (2008) Information Sharing and Exchange in the Context of Product Lifecycle Management: Role of Standards. *Computer-Aided Design*. 40(7), 789–800.

Richards, D. (2006) Ad-Hoc and Personal Ontologies: A Prototyping Approach to Ontology Engineering. In A. Hoffmann, B. Kang, D. Richards & S. Tsumoto (Eds.), *Advances in Knowledge Acquisition and Management: PKAW 2006: Lecture Notes in Computer Science*. Vol. 4303. Berlin, Heidelberg: Springer.

Roehner, N., Beal, J., Clancy, K., Bartley, B., Misirli, G., Grünberg, R., Oberortner, E., Pocock, M., Bissell, M., Madsen, C. & Nguyen, T. (2016) Sharing Structure and Function in Biological Design with SBOL 2.0. *ACS Synthetic Biology*. 5(6), 498–506.

Schulz, M., Krause, F., Le Novère, N., Klipp, E. & Liebermeister, W. (2011) Retrieval, Alignment, and Clustering of Computational Models Based on Semantic Annotations. *Molecular Systems Biology*. 7(1), 512.

Schyfter, P. (2012) Technological Biology? Things and Kinds in Synthetic Biology. *Biology and Philosophy*. 27, 29–48.

Smith, B., Ashburner, M., Rosse, C., Bard, J., Bug, W., Ceusters, W., Goldberg, L. J., Eilbeck, K., Ireland, A., Mungall, C. J. & Leontis, N. (2007) The OBO Foundry: Coordinated Evolution of Ontologies to Support Biomedical Data Integration. *Nature Biotechnology*. 25(11), 1251.

Sprinzak, D. & Elowitz, M. (2005) Reconstruction of Genetic Circuits. *Nature*. 438(7067), 443–448.

Swanton, C. (2003) *Virtue Ethics: A Pluralistic View*. Oxford: Clarendon Press.

Tamames, J., Moya, A. & Valencia, A. (2007) Modular Organization in the Reductive Evolution of Protein-Protein Interaction Networks. *Genome Biology*. 8(5), R94.

Wilkinson, M., Dumontier, M., Aalbersberg, I., Appleton, G., Axton, M., Baak, A., Blomberg, N., Boiten, J. W., da Silva Santos, L., Bourne, P. & Bouwman, J. (2016) The FAIR Guiding Principles for Scientific Data Management and Stewardship. *Scientific Data*. 3.

Wimsatt, W. (2007) *Re-Engineering Philosophy for Limited Beings*. Cambridge: Harvard University Press.

Wong, D. L. & Baker, C. M. (1988) Pain in Children: Comparison of Assessment Scales. *Pediatric Nursing*. 14(1), 9–17.

Zundel, Z., Samineni, M., Zhang, Z. & Myers, C. J. (2017) A Validator and Converter for the Synthetic Biology Open Language. *ACS Synthetic Biology*. 6(7), 1161–1168.

Zúñiga, G. L. (2001) Ontology: Its Transformation from Philosophy to Information Systems. In *Proceedings of the International Conference on Formal Ontology in Information Systems*. Vol. 2001 (pp. 187–197). New York: ACM Press.

Scientific, philosophical, and legal challenges in classifying biological constructs

6 What is a new object?

Case studies of classification problems and practices at the intersection of law and biotechnology

Ubaka Ogbogu

Introduction

Biotechnology utilizes biological objects to create new objects that may vary in constitution, character, or purpose from existing objects and kinds. Advances in biotechnology, such as the creation of synthetic organisms (Zhang et al., 2017), genome editing via CRISPR (Xiao-Jie et al., 2015), and somatic cell nuclear transfer (cloning) (Wilmut et al., 2002), have produced objects that challenge our thinking and knowledge about kinds and classes in biology and physiology. Often, this challenge is met in the natural sciences by allocating new objects to taxonomies, established or new, that are defined by similar (physical) characteristics, and grounded, as much as possible, in testable and verifiable facts derived from knowledge about our natural world (Khalidi, 2013, Chapter 2; Samuel, 2005, p. 284; Takhtajan, 1973, p. 17). As Khalidi (2013, p. 42) observes, "Scientific categories aim at identifying the kinds that really exist in the world, as opposed to inventing categories that will merely serve certain parochial practical ends".

A more vexing challenge arises when the products of biotechnology are introduced to society, and thereby encounter opinions, values, discourses, and classificatory schemes that are not founded on the scientific method. For example, the announcement of the creation of the first cloned mammal (the famous Dolly the Sheep) did not cause as much of a stir in the world of scientific classification as it did among societal observers, who sought to understand and classify the object through lenses beyond its physical characteristics, including its social implications (such as fears that clones will be classified and treated differently, as inferior entities, or used to serve the ends of others) (Wade, 2013). Similarly, the genetically engineered bacterium *Pseudomonas putida* shares a scientific genus with numerous naturally occurring Gammaproteobacteria but was at the centre of legal disputes regarding the patentability of living organisms and the dual-use implications of higher life forms (*Diamond v. Chakrabarty*, 1980).

The architects and practitioners of the law are often summoned to make pronouncements on or adjudicate social disputes regarding the appropriateness, relevance, and consequences of biotechnological inventions. The involvement of law and the legal system often entails the classification of such inventions, mainly as a means of determining whether they fit into existing regulatory categories or require novel or special regulatory treatment. Such classification, in turn, involves the making or remaking of scientific and social order in ways that impact the scientific enterprise, and that can cohere or conflict with the corresponding role of science in "securing social stability and order" (Jasanoff, 2008, p. 761).

How, then, do law and legal systems treat or classify new objects in biotechnology, especially objects that present as deviations from existing or conventional objects and kinds? This chapter seeks to answer this question through an analysis of three case studies of classification practices and problems at the intersection of law and biotechnology. The first case study deals with classifications of human embryos when distinguishing between legally legitimate science and criminal conduct. The second focuses on classifications that are used to delineate the legal fate of excised human tissue for the purpose of assigning property and related legal rights, and the consequences of such classifications for scientific research. The third and final case study deals with classifications applied to transgenic animals, genetically modified organisms, and other biotechnological inventions, for establishing whether a patentable invention exists. The case studies draw on and analyze legal rules and doctrines from Canada, the U.K., and the United States, as they relate to these classification problems.

A common theme that emerges from these diverse case studies is that classification is a means through which law wields authority over scientific discourse and activities. This authority can be, but is not necessarily, antagonistic to science and its aims. Rather, it often produces collaboration between law and science in defining and regulating social order and progress. Through the case studies, the chapter also aims to provide insights into approaches to legal and scientific classification in relation to matters at the law-science interface, the degree of congruence between classification approaches and outcomes (i.e., whether legal and scientific classifications are aligned or misaligned), and the impacts and consequences of legal classifications on scientific aims, methods, processes, and outcomes, especially as they relate to biotechnology.

The case studies, which make up the principal sections of the chapter, are bookended by a context-setting overview of the state and practice of classification in the field of law, which canvasses the history, objects, methods, purposes, and outcomes of legal classification, and a concluding section that summarizes and reiterates the main points and lessons that emerge from the case studies.

Law and classification

Unlike the natural sciences, where a rich history of classification exists, and where taxonomies are a primary means of organizing knowledge, lawyers rarely evince or "experience more than an oblique concern for classification" (Jamieson, 1988, p. 552). There is some interest in and introspection on the subject in continental legal systems based on the civil law tradition, which are organized around, and adhere strictly to, taxonomies that define and regulate sources of law (case law and legislation), areas of legal relations (public and private law), and the objects of legal rules (persons, things, and actions) (Gaius, 1908, 1st Commentary, Title II; Samuel, 2005). Classification, in theory and practice, has been largely ignored in the other major Western legal tradition or system, namely the common law.[1] As Samuel (2005, p. 272) observes, "Little sophisticated work has been done on classification and its relationship with legal reasoning" in the common law world.

The foregoing suggests that legal rules and concepts lack objects that require classification, or the absence of legal classification methods, purposes, and outcomes. On the contrary, law and legal work is based on systems of classification that are used to define, persuade, interpret, and/or organize the objects of law, including doctrine (binding judicial decisions), legal reasoning, statutory rules, or even elements of the natural world (persons, things) that the law regulates or protects. Feinman (1989, p. 664) notes that classification is an "important and ubiquitous aspect of legal reasoning". Legal classification, like classification in other disciplines, serves to impose conceptual precision and order on the objects of law (by sorting them into clear categories based on shared characteristics, history or effect, for example), to introduce the naïve and initiated to the episteme, customs, and traditions of the discipline, and in a less obvious or directed way, to provide an internal map of modes of reasoning and ordering within a discipline that affects and regulates social order in fundamental ways. This internal map can also provide a basis for assessing congruence with classification systems in other institutions responsible for making and regulating social order, such as science, politics, and religion. It is in the latter sense that the law has the most purchase in its encounter with biotechnology (and science generally). If scientific objects have any real-world relevance or application, they must also exist as legitimate legal objects, or fall into a lacuna that the law is agnostic about.

Among the reasons for the tenuous interest in classification among lawyers, one stands out in relation to the focus of this chapter. Law, it is claimed, is inherently complex, and consists of doctrines, rules, and concepts (collectively, objects) that are difficult to categorize or classify because they do not function or operate in isolation from each other (Jamieson, 1988; Samuel, 2005; Waddams, 2003). Legal knowledge and objects, therefore, are cumulative rather than exclusive, and tend to arise from diverse influences that are difficult to compartmentalize. For example, a tort, or civil wrong, is not a singular but a complex concept that is derived from, subsists in, and interacts with diverse influences, including contracts, property, and public policy. Though developed largely to govern the interactions between private citizens rather than the relationship between citizens and the state, several torts, such as negligence and nuisance, have both public and private dimensions. Such torts, therefore, have fluid boundaries that defy a classificatory divide between private and public law. Furthermore, legal objects are not static, but evolve with society. As Baker (2000) notes, legal history is about legal change, and such change constrains classification, or at a minimum, creates a need for an evolving system of classification. While this is not necessarily a bar to the practice of classification itself, it is a daunting proposition in an institution with high tolerance for "practical convenience" and "chaos", and which over-relies on fictions and terms of art as instruments of social change (Jamieson, 1988).

If law lacks a culture of consistent or systematic classification and categorization of objects, then one can safely assume that the field and its practitioners do not consciously engage with extra-legal objects through a classificatory lens.[2] To the extent that any engagement exists, it is likely to reference available or existing internal traditions of classification rather than import or recognize the classification systems or practices that attach to extra-legal objects. Thus, as law engages with a scientific object, whether in legislation or a legal dispute, the impulse to discover or establish the object's essence or extra-legal niche is neither a primary nor an essential purpose of the engagement. A scientific object that enters the legal process may, therefore, emerge from the process intact, or, tainted and distorted (in helpful or harmful ways) by meanings and interpretations imposed by a field oblivious to its scientific order and associated meanings. Indeed, as legal scholar Annalise Acorn observes (personal communication, October 16, 2017), it is not uncommon for law to "impose a class (or some other form of meaning) on the external world, as an exercise of power, without any sensitivity to shared reality [or reality *per se*]".

The case studies that follow illuminate the latter point and discuss the implications for both law and science as both institutions engage with making and directing social order.

What is an embryo?

The first case study is drawn from *Regina v. Secretary of State for Health ex parte Quintavalle* (*"Quintavalle"*), a 2003 U.K. House of Lords decision. The case focussed on a dispute between the petitioner, Josephine Quintavalle, on behalf of Pro-Life Alliance, a U.K. organization that advocates for legislation to protect the unborn from the point of conception, and the U.K. Secretary of State for Health, regarding whether a human clone should be legally classified as an embryo under U.K. law. A human clone is created by using nuclear material extracted from a human egg to reprogramme an enucleated human adult cell (i.e., a cell that has had its nucleus removed). At the time, the term "embryo" was legally defined as "a live human embryo where fertilisation is complete" (Human Fertilisation and Embryology Act of 1990). In literal terms, this definition anticipated two main characteristics of the defined object or class, namely that the term (a) refers to a live human organism conceived *in vitro*, and (b) requires that creation of the organism occur through the process of fertilization (the union of human egg and sperm). The petitioner claimed that this definition excludes novel human organisms created by means other than fertilization or, stated differently, that such organisms are outside of the regulated class. Consequently, the U.K. regulator was not authorized to permit the creation of organisms that fall outside of the regulated class. If the court accepted this claim, the effect would be to place the clone (and other novel human organisms) in a regulatory lacuna, as the creation of human organisms in a manner or through processes that do not conform to law is proscribed by law.

In developmental biology – the scientific field primarily engaged with the study of developing organisms – the term "embryo" describes a class of organisms in the earliest stages of development (Guenin, 2003; see also Chapter 7). While these organisms may have different names, such as clones or hybrids (an embryo created by combining human and animal reproductive cells), they belong to the class *embryo*. Differences within the class diminish further when viewed through the lens of intended purpose for creation. The primary purposes for creating and utilizing conventional and non-conventional embryos are generally the same, namely to conduct research studies into early human development, to improve fertility treatments, and to develop cell-based therapies and testing assays from human pluripotent stem cells derived from the embryos (American Academy of Pediatrics, 2001; National Institutes of Health, 2016). The class *embryo* is, therefore, conceptually fluid, and it encompasses a variety of organisms that are unified by purpose and (some) class characteristics.

The petitioner's position in *Quintavalle*, therefore, posed serious challenges for scientific research involving human embryos. The dispute arose at a time when U.K. scientists sought to create human embryos from non-conventional elements (e.g., a combination of reproductive and non-reproductive human cells) and/or through non-conventional processes (e.g., nuclear reprogramming triggered by electric shock). The non-conventional embryos, including clones and hybrids, were intended to serve as assays for scientific studies of developmental biology and regenerative medicine, and as a substitute for conventional human embryos, research uses of which are socially controversial (Cavaliere, 2017; Hurlbut, 2017; Wertz, 2002).

In its ruling on the case, the House of Lords emphasized the purposive unity of the class. The court acknowledged the "obvious attraction" of a literal reading of the definition of the term "embryo" in the governing legislation but opted for a "liberal and permissive" (*Quintavalle*, paras. 14–15) interpretation that adheres to the scientific understanding of the term embryo as including a variety of organisms that are "similar" in "kind or dimension", and which treats cloned human embryos as "another member of the same genus" (*Quintavalle*, paras. 15–16). The members of the court offered a number of reasons to support this interpretation. Lord Bingham of Cornhill reasoned that cloned human embryos (and other non-conventional embryos) were not a scientific possibility when the law was enacted (i.e., they did not exist at the time of enactment) (*Quintavalle*, paras. 1–19). Therefore, the legislator could not have employed a definition of the term that included non-existent entities as part of a known class. By implication, if cloned human embryos existed at the time of enactment of the law, they would surely have been addressed in the definition because they "fall within the same genus of facts as those which the expressed [legislated] policy" advances (*Quintavalle*, para. 15). Lord Steyn similarly favoured an "updating interpretation" that brings novel scientific developments within a prior regulatory class if permitted by the purpose(s) for which the law was enacted (paras. 20–36). He also endorsed the purposive approach to interpretation, which eschews a literal reading of words used in legislation in favour of a reading that examines and promotes the purpose or object of the legislation[3] (para. 21).

Quintavalle demonstrates collaboration between law and science in achieving and fostering correspondence between legal and scientific categories (and discourses). The court's treatment of the disputed question not only displays a deference to scientific epistemology but also shows that methods of legal classification are robust and nimble enough to accommodate classificatory practices in science and other fields. Furthermore, the court's reference to the similarity of all members of the class in "kind and dimension" suggests that the legal class is defined not just by the internal attributes or characteristics of the genus or kind, but also by functional or practical purpose(s). Viewed in the latter sense, character and purpose become elements that define and encompass the class, thus advancing unity of the class that may not be achieved by emphasizing one without, or over, the other.

This unifying approach to interpretation (and classification) taken by the House of Lords has been rejected or does not obtain in other jurisdictions. Canadian law defines the term "embryo" as both an organism in the conventional sense and a class that includes novel organisms that are different from the embryo, such as a human clone and a hybrid (Assisted Human Reproduction Act of 2004). An embryo is defined as a

> human organism during the first 56 days of its development following fertilization or creation, excluding any time during which its development has been suspended, and includ[ing] any cell derived from such an organism that is used for the purpose of creating a human being.
>
> (Assisted Human Reproduction Act of 2004, § 3)

A human clone is an embryo that is created by manipulating human reproductive material obtained from a single human being, foetus or embryo (Assisted Human Reproduction Act of 2004, § 3). A human clone differs from a conventional embryo because the latter is created by combining human male and female reproductive materials. Further, a hybrid is defined to include interspecies embryos created by combining human and animal reproductive materials (Assisted Human Reproduction Act of 2004, § 3).

In Canada, different legal obligations and consequences attach to embryos, clones, and hybrids. The creation of embryos and hybrids is permitted by law, while the creation of a clone is prohibited (Assisted Human Reproduction Act of 2004, § 5). Hybrids – but not embryos – can be created specifically for research purposes (Assisted Human Reproduction Act of 2004, § 5). Embryos can be implanted into a woman, but hybrids and clones cannot (Assisted Human Reproduction Act of 2004, § 5). Canada's highest court, the Supreme Court, has endorsed the classification and associated prohibitions, on the basis that human cloning and the creation of entities with human and animal characteristics "raise moral concerns long before such experiments result in the creation of a new life form" (*Reference re Assisted Human Reproduction Act*, 2010, para. 108). The application of a moral lens to the classification of members of the same (biological) genus imports a non-scientific, non-epistemic, and non-factual describer into the classification.

The Canadian approach suggests a difference between scientific and legal classification cultures. Legal reasoning, unlike science, is not founded on exclusive describers, but on a whole range of describers and an interaction of describers (Samuel, 2005). By treating some members of the class as moral outliers that attract legal proscription, the Canadian approach creates two legal realities, one in which some members of the class (conventional embryos, hybrids) are regulated objects, while others (clones) are legally prohibited. Character and scientific purpose matter less in this alternate reality than moral valuation. Since classification based on moral criteria is likely to lack the "correspondence between fact and scientific discourse" that is a "vital epistemological test of validity" (Samuel, 2005, p. 284) in the natural sciences, the approach to classification exemplified by Canadian law destabilizes or upends the scientific approach to classification.

The possibility of creating a human embryo using gametes derived from adult human cells (e.g., a skin cell) complicates this legal classification exercise. In recent groundbreaking studies, scientists genetically reprogrammed adult cells derived from a single human donor to produce progenitor cells with characteristics similar to the cells found in the developing human embryo (Takahashi et al., 2007). Other studies have shown that it may be possible in the future to derive sperm and ova from these genetically reprogrammed cells (Eguizabal et al., 2011; Hayashi et al., 2011; Okita et al., 2007). If the derived gametes are combined, the result would be a human embryo with a diploid set of chromosomes derived from a single donor. In the U.K., the resulting organism would be classified as an embryo, and its creation and use for research studies is permissible by law. By contrast, Canadian law classifies such organism as a clone, and creation or use for any purpose will attract criminal sanction. The latter effect is the unintended consequence of a broad classification that groups single-source-derived human organisms differently from organisms created by conventional fertilization, regardless of physical similarities or characteristics (Rugg-Gunn et al., 2009; Ogbogu and Rugg-Gunn, 2008). It is unintended because adult-cell-derived human organisms were not a scientific possibility at the time the Canadian law was enacted, and could, therefore, not have been within the lawmakers' contemplation in creating the class. However, as *Quintavalle* suggests, it is not uncommon for the law to create or enable a class that includes known and future (unknown) objects, and that is bounded by legal reasoning rather than reality or monophyletic characteristics.

Altered excised human tissue as a new object

Histopathologists classify and categorize excised human tissue and other biological material based on the biological properties (i.e., morphology, structure, and molecular characteristics) of the material and its associated derivates (Mills et al., 2015; Bemmels et al., 2012). Law, by contrast, classifies such material, at first instance, by the purpose of or reason for excision, mainly for clinical or research use (Cheung et al., 2013). The different modes of classification are consistent with the manner through which both institutions interact with and direct natural and social order. The aim of the clinical or scientific classification is to impose coherence on objects by ordering their immanent rather than relational characteristics.

The legal classification is not concerned with internal attributes, but instead seeks to understand the relations between the objects and their social utility. In short, the different systems of classification are consistent with the culture and aims of each institution.

However, both classification systems align in the view that an unaltered excision is an existing object, while an altered excision is a new object (or at least a new form of the existing object). In histopathologic terms, the distinction between both objects (or forms of the object) has bearing mainly on practical uses or applications; unaltered excisions (archived or excess) can be used for diagnostic, research, teaching or quality assurance purposes, but once altered, can no longer be used for diagnostic or related clinical purposes (Ogbogu and Mengel, 2013). In legal terms, the distinction also matters in a different but significant respect, namely the allocation of property rights over the object. This allocation, in turn, complements and advances scientific aims, while significantly diminishing the legal rights of the person from whom the tissue is excised.

The landmark decision of the Supreme Court of California in *Moore v. Regents of the University of California* ("*Moore*") (1990) is illustrative. The case dealt with the issue of whether a person has property rights over his or her own biological material excised during a clinical procedure. The plaintiff in the case, John Moore, received treatment for hairy-cell leukaemia, a form of cancer, from the Medical Center of the University of California at Los Angeles (UCLA). During the treatment, biological materials, including Moore's spleen, were removed from his body for clinical reasons. Subsequently, and without Moore's knowledge or consent, the physicians attending to him used part of Moore's spleen for research activities. The research led to the development of a cell line, which was patented by UCLA. Moore's physicians, named as inventors on the patent, were joint beneficiaries with UCLA of any royalties and profits arising out of the patent. Commercialization of the cell line soon followed, resulting in financial gains to UCLA and the physicians.

Moore, upon discovering what transpired with his biological material, sued the physicians and UCLA. He claimed that the actions of the physicians and UCLA violated his ownership rights over the tissue and resulting cell line, and amounted to conversion, which is an unlawful act that interferes with possession or ownership of personal property. Conversion involves taking and using another's property in a manner that is inconsistent with that other's right to possess or own the property, and without consent or authorization (Osborne, 2003, p. 277). A common example of conversion is stealing someone else's property. According to Moore, the unauthorized use of his cells amounted to conversion. Since proof of conversion depends on proving a right to possess or own the converted object, Moore was asserting a proprietary (i.e., possessory or ownership) interest in the excised spleen, and/or in the cell line and products developed from it.

Though not argued explicitly in the case, the asserted proprietary interest originates or derives from Moore's right over his body (i.e., the right we have as individuals over our living bodies and our physical integrity). Excised tissue (especially in an unaltered form) is, in a sense, an extension, albeit detached, of an existing object (i.e., Moore's body), and thus, an existing object *per se*. If the law of conversion protects Moore's bodily integrity, does that protection extend to tissue excised from that body, or to materials derived from the tissue? At what point, if at all, do derivates become new objects and thereby lose constitutional or juridical connection to the original source?

In its judgement, a majority of the court ruled that there was no basis in law to find that Moore retained a proprietary interest in his excised biological materials. By extension, he also could not claim a legally supportable proprietary interest over materials and products derived from the excised biological materials. In the court's view, the pendulum may even swing in the opposite direction, because California law dealing with the disposal and handling of excised biological materials "drastically limits any continuing interest of a patient in excised cells" (*Moore*, p. 137). Without clear and direct legal authority to support ownership or possession post excision, there could be no basis "for importing the law of conversion" into the matter at hand (*Moore*, p. 137). In sum, Moore had neither ownership nor possessory interests in the excised biological materials and associated derivates, regardless of the reason for, or purpose of, excision.

The court's decision suggests that, from a legal standpoint, excised tissue is a new legal object for purposes of assigning legal rights in a property law context. The biological properties of the object, and the reasons for or purpose of excision, are not relevant considerations in determining the vesting of rights *in rem*. Rather, the mere fact of excision generates a change in the legal classification and fate of the object. In this transformed state, the excised tissue no longer belongs to the person from whom it is excised, and reversionary and other property-related rights are completely extinguished.[4]

If Moore does not own tissue excised from his body, then who does? While the court avoided comment on this intriguing question by stating that an answer was not necessary to the disposition of the case, it found that the tissue, if altered post excision, is owned by the person whose effort or ingenuity produced the altered object. The court, in effect, recognized two classes of excised tissue for purposes of assigning property rights: unaltered versus altered excisions. Unaltered excisions are not subject to property rights (and are not owned by the patient), while altered excisions, as new objects, are subject to property law principles, and are owned by the person who has rights over the altered form, because the alteration is a "product of human ingenuity" (*Moore*, p. 142).

It is unclear from *Moore* what amount or degree of alteration will trigger a reclassification of excised tissue. However, doctrine from patent cases – where a similar construction of "products of human ingenuity" is used to classify and distinguish patentable from unpatentable kinds – suggests that the alteration should result in an object that is "markedly different" from the natural form of the object (*Diamond v. Chakrabarty*, 1980; *In re Roslin Institute* (*Edinburgh*), 2014). This construction, which is inconsistent with the result in *Moore*, will disqualify unaltered derivates, such as cell lines and nucleic acids, and derived data, such as genetic, cytogenic, and epigenetic information, from the propertied class. An alternate view is that alteration refers to any application of human ingenuity

regardless of a change in the physical properties of the object. Thus, the alteration, and the consequent reclassification, are legal fictions that are provable not by
reference to physical or shared attributes of the object or class, but by reference
to the legal class or category itself (i.e., product of human ingenuity). Viewed in
this sense, the legal class defines and constitutes, a priori, the facts and objects
that make up the class. This insight would likely baffle the natural scientist, who
is accustomed to constructing classes from observed facts.

The rationale for treating tissue excisions and human-derived alterations as new
objects for purposes of assigning property rights is not just a matter of legal doctrine (or the absence of it). The court's opinion in *Moore* suggests that the treatment
is also rooted in the concern that imposing liability on scientists for appropriating
human biological materials for research purposes would have stifling effects on
"socially important" and "lucrative" science (*Moore*, p. 135). As one judge in the
case observed, the court simply could not allow individuals to claim "ownership of
the results of socially important medical research" because doing so "would affect
medical research of importance to all of society" (*Moore*, p. 135). Putting aside
some obvious problems with this reasoning – such as the absence of any explanation as to why appropriation for the sake of science is necessary or legitimate, and
the absence of proffers of proof that anyone other than the parties sued in the case,
let alone society, stood to benefit from the appropriation – the majority opinion can
be characterized as pro-science. The court's decision, in effect, takes judicial notice
of, and treats as self-evident, the socio-economic value of scientific research, albeit
grounded on untested assumptions, and on a classificatory approach that renders
excised biological materials a different legal object for purposes of assigning property rights.

Is a genetically engineered organism a new object?

A third and final case study relates to classifications applied to human-made
organisms, for purposes of assigning patent rights. A patent is a legal right that
allows the holder to exclusively make, use, or sell an invention, and to exclude
others from making, using, or selling the invention, for a given period. A patentable invention is, generally, a thing or process made, manufactured, or discovered,
that is new, useful, and non-obvious to a knowledgeable observer or assessor.
While this definition is broad enough to accommodate natural and human-made
objects and phenomena, things found in nature, or "products of nature", including
"the laws of nature [and] physical phenomena" are not patentable subject matter
(*Diamond v. Chakrabarty*, 1980, pp. 308–309). Thus, plants, animals, and other
genera that exist in nature are not patentable, regardless of whether they ordinarily meet the definition of an invention, that is, they are discovered, new, useful,
and non-obvious to a knowledgeable person. This way, patent law recognizes two
sets of binary classifications for purposes of assigning or denying patent rights,
namely new versus existing objects, and natural versus human-made objects.

The latter classification has proved contentious in the context of patent claims
for human-made, "non-naturally occurring" (or "artificially bred") organisms,

especially where such organisms exhibit "markedly different characteristics from any found in nature" (*Diamond v. Chakrabarty*, 1980, p. 310). The issues that arise in these claims are twofold. First, whether the organism, *qua* organism, is a natural thing or object, or nature-like, and therefore unpatentable (even if new). Second, whether the artificial characteristics of the organism qualify it to be classified as an unnatural thing or object, and therefore patentable. Both matters were considered in *Diamond v. Chakrabarty* ("*Chakrabarty*"), a seminal 1980 decision of the Supreme Court of the United States, which dealt with the patentability of a live, human-made micro-organism, a genetically modified transgenic bacterium derived from the genus *Pseudomonas* (which includes over 100 naturally occurring species). The bacterium was invented by Ananda Mohan Chakrabarty, a microbiologist employed by General Electric. Unlike its naturally occurring counterparts, the bacterium can break down crude oil, and is, therefore, useful for remediation of environmental disasters caused by oil spills. General Electric sought to patent the invention, but the application was rejected by the patent examiner because the bacterium was a "product of nature" and "living thing", and as such, unpatentable.

The Supreme Court disagreed and reversed the examiner's decision. A majority of the Court classified the bacterium as an unnatural thing, noting that it was "a nonnaturally occurring manufacture or composition of matter . . . with markedly different characteristics from any found in nature" (*Chakrabarty*, pp. 309–310). In the Court's view, therefore, the bacterium belongs to a different genus, or at a minimum, to a species or kind within the genus defined by non-natural occurrence and functional characteristics introduced by human intervention (i.e., the ability to break down oil spills). This interpretation seems inconsistent with scientific approaches to bacterial taxonomy, which are based (broadly) on phenotypic, chemotaxonomic, and genotypic, rather than functional, classification systems (Schleifer, 2009). Furthermore, Chakrabarty's bacterium belonged to a class that was defined not just by reference to non-natural occurrence and *functional* human-enabled characteristics, but also by the fact that it was a product of human ingenuity.

Thus, consistent with the approach in *Moore*, the Supreme Court acknowledged and endorsed human agency as a defining characteristic of a legal object or kind, and as a characteristic that justifies a departure from legal precedent. Similar to *Moore*, the departure favours, advances, or facilitates a scientific goal, outcome, or claim. In *Moore*, the resulting claim is that medical research is a social good that should be advanced and bolstered by legal reasoning. Likewise, the reasoning in *Chakrabarty* treats human ingenuity as a desirable and protected social good and enables a liberal construction of patent rights that is favourable to the claims of the human inventor. Indeed, *Chakrabarty* is often cited as a fundamental shift in U.S. patent law, and as enabling the most liberal patent regime worldwide. As Chief Justice Burger noted in *Chakrabarty* (quoting Congressional Committee Reports accompanying U.S. patent law – Title 35 of the United States Code), patentable subject matter includes "anything under the sun that is made by man" (*Chakrabarty*, p. 309).

This expansive reading of patent law, and the classificatory approach that made it possible, was rejected by the Canadian Supreme Court in *Harvard College v. Canada (Commissioner of Patents)* ("*Harvard College*"), a 2002 decision of the Court that dealt with the patentability of transgenic animals. The case arose when the Commissioner of Patents refused to issue a product patent for genetically altered mice developed by Harvard University researchers. The purpose of the genetic alteration was to make the mice more susceptible to carcinogens, and therefore more useful for cancer research. Harvard, in their petition to the Court, urged consideration similar to the U.S., where a patent had been issued for the "oncomouse", no doubt as a consequence of the decision of the United States Supreme Court in *Chakrabarty*. Refusing to be drawn into "[c]omparisons with the patenting schemes of other countries" and debates regarding whether "higher life forms such as the oncomouse ought to be patentable", the Court concluded that the oncomouse was not patentable because it did not satisfy the definition of an invention under Canadian law (*Harvard* College, p. 47).

The Court reached this conclusion by engaging in a classification exercise, albeit one that favoured a more restrictive interpretation of the meaning of the term "invention" under Canadian law. According to the Court, an invention is defined in the governing statute as either a "manufacture" or "composition of matter", and not [as Justice Burger supposed] "anything new and useful made by man" (*Harvard College*, p. 47). A higher life form, such as the oncomouse, is not a "manufacture" because the term denotes "a non-living mechanistic product or process" (*Harvard* College, p. 47). While a "composition of matter" could (and does indeed) include living things, the Court held that higher life forms are not compositions of matter in the sense intended by the Patent Act. A summary of the ruling (contained in the headnotes) suggests a classificatory sleight of hand, and bears reproduction in full:

> The words "composition of matter" . . . as they are used in the Act do not include a higher life form such as the oncomouse. The words occur in the phrase "art, process, machine, manufacture or composition of matter". A collective term that completes an enumeration is often restricted to the same genus as the terms which precede it, even though the collective term may ordinarily have a much broader meaning. While a fertilized egg injected with an oncogene may be a mixture of various ingredients, the body of a mouse does not consist of ingredients or substances that have been combined or mixed together by a person. . . . Higher life forms cannot be conceptualized as mere "compositions of matter" within the context of the *Patent Act*.
>
> (*Harvard College*, pp. 47–48)

At least two classification schemes are apparent in the quoted excerpt. First, the Court classifies "composition of matter" as belonging to a terminological genus that excludes living things (or natural kinds). Thus, even if the oncomouse is ordinarily a living thing and, by extension, a composition of matter, it is not a composition of matter in the limited sense allowed by the terminological genus. Differently stated, the oncomouse may be a composition of matter in a general sense, but not in the particular sense allowed by the class of terms used in the statute. Here, the Court differs from the inclusive approach in *Chakrabarty*, where the term "invention" was construed as including natural and unnatural objects. Second, the Court classified the oncomouse as purely a natural object (and thus outside the scope of composition of matter in the *Patent Act*) because the genetic modifications in the oncomouse are natural, rather than functional, elements, as was the case with Chakrabarty's bacterium.

Conclusion

The three case studies discussed in this chapter provide distinct perspectives on how law engages and interacts with new objects produced by biotechnology. The first case study, which deals with the classification of novel, unconventional embryos for purposes of attaching legal consequences to their creation and handling, suggests that such new objects fare better in a legal classification scheme if viewed as similar to existing legal objects. Hence, novelty exists in two distinct legal senses: as a manifestation of legal reality, and as an unknown entity that attracts novel (and often antagonistic) scrutiny and regulation. The second case study relies on legal disputes over ownership of excised human tissue to show that novelty, in legal terms, can be a construct or artefact that is devoid of any immanent, physical or biological characteristics. In this context, novelty is constructed as an exception to established legal rights, interests, boundaries, and consequences. The final case study, which examined the legal concept of an invention for patenting purposes, highlights how legal classifications of new objects can advance or impede scientific aims. Together, these engagements highlight how law shapes and influences biotechnology (and science more broadly) in important ways and mediates the tensions and opportunities that arise when biotechnology confronts social forces.

Acknowledgements

I am grateful to those who worked on the paper with me: my wonderful research assistants Jenny Khakh, Jenny Du, and Yonida Koukio; my colleagues, Professors Annalise Acorn, Tamara Buckwold, and Anna Lund, who reviewed drafts and provided comments and suggestions; and Julia R. S. Bursten, Evan Hepler-Smith, and other participants in the Unnatural Kinds workshop.

Notes

1 The tradition that supplied the roots of English, American, Canadian, Australian, and "Commonwealth" law. The defining features of the common law are judge-made rules and the practice of following precedents, that is, applying existing judicial decisions to resolve new cases (disputes) that have similar facts. By contrast, the civil law system is primarily based on a codified set of rules created by a legislative body.
2 Or, as Julia Bursten observed in a review of this chapter, because legal objects rely on norms of individuation that defy the classification practices suggested by classification debates in the epistemology and the sciences, it would be surprising to find that lawyers

and legal scholars tacitly endorse those classification practices anywhere in their professional lives.

3 This approach is embodied in the injunction issued by Justice Learned Hand in *Cabell v. Markham* (1945) "not to make a fortress out of the dictionary; but to remember that statutes always have some purpose or object to accomplish".

4 The legal connection between a person and her excised tissues is not entirely broken. The court found that the physician-scientists violated Moore's consent rights, as well as fiduciary duties owed to him, by deviating from the reasons for, or purpose of, excision, without Moore's consent or knowledge. Since Moore authorized the removal and use of the biological materials only for clinical purposes, the materials could only be used for that purpose, and could only be assigned to another purpose with Moore's consent.

References

American Academy of Pediatrics. (2001) Human Embryo Research. *Pediatrics.* 108(3), 813–816.

Assisted Human Reproduction Act, Statutes of Canada. (2004) Chap. 2. Retrieved from http://laws-lois.justice.gc.ca/.

Baker, J. H. (2000) Why the History of English Law Has Not Been Finished. *Cambridge Law Journal.* 59(1), 62–84.

Bemmels, H. R., Wolf, S. M. & Van Ness, B. (2012) Mapping the Inputs, Analyses, and Outputs of Biobank Research Systems to Identify Sources of Incidental Findings and Individual Research Results for Potential Return to Participants. *Genetics in Medicine.* 14(4), 385–392.

Cabell v. Markham. (1945) 148 F.2d 737.

Cavaliere, G. (2017) A 14-Day Limit for Bioethics: The Debate over Human Embryo Research. *BMC Medical Ethics.* 18(1), 38–49.

Cheung, C. C., Martin, B. R. & Asa, S. L. (2013) Defining Diagnostic Tissue in the Era of Personalized Medicine. *Canadian Medical Association Journal.* 185(2), 135–139.

Diamond v. Chakrabarty. (1980) 447 U.S. 303.

Eguizabal, C., Montserrat, N., Vassena, R., Barragan, M., Garreta, E., Garcia-Quevedo, L., Vidal, F., Giorgetti, A., Veiga, A. & Izpisua Belmonte, J. C. (2011) Complete Meiosis from Human Induced Pluripotent Stem Cells. *Stem Cells.* 29(8), 1186–1195.

Feinman, J. M. (1989) The Jurisprudence of Classification. *Stanford Law Review.* 41(3), 661–717.

Gaius. (1908) The Institutes. In J. Trapnell & B. A. Graham (Trans.), *The Making of Modern Law: Foreign, Comparative and International Law,* 2012. Retrieved from http://galenet.galegroup.com/servlet/MMLF?af=RN&ae=HT100326966&srchtp=a&ste=14&locID=edmo69826.

Guenin, L. M. (2003) On Classifying the Developing Organism. *Connecticut Law Review.* 36, 1115–1131.

Harvard College v. Canada (Commissioner of Patents), 2002 SCC 76. Retrieved from https://scc-csc.lexum.com/scc-csc/en/nav.do.

Hayashi, K., Ohta, H., Kurimoto, K., Aramaki, S. & Saitou, M. (2011) Reconstitution of the Mouse Germ Cell Specification Pathway in Culture by Pluripotent Stem Cells. *Cell.* 146(4), 519–532.

Hurlbut, J. B. (2017) *Experiments in Democracy: Human Embryo Research and the Politics of Bioethics.* New York: Columbia University Press.

In re Roslin Institute (Edinburgh), 750 F.3d 1333. (2014) Retrieved from https://scholar.google.com/.

Jamieson, N. J. (1988) Legal Classification and the Science of Law. *Otago Law Review.* 6(4), 550–562.

Jasanoff, S. (2008) Making Order: Law and Science in Action. In E. J. Hackett, O. Amsterdamska, M. Lynch & J. Wajcman (Eds.), *The Handbook of Science and Technology Studies* (pp. 761–786). Cambridge: MIT Press.

Khalidi, M. A. (2013) *Natural Categories and Human Kinds: Classification in the Natural and Social Sciences.* New York: Cambridge University Press.

Mills, S. E., Greenson, J. K., Hornick, J. L., Longacre, T. A. & Reuter, V. E. (Eds.). (2015) *Sternberg's Diagnostic Surgical Pathology.* Philadelphia: Lippincott Williams & Wilkins.

Moore v. Regents of the University of California, 51 Cal.3d 120. (1990) Retrieved from https://scholar.google.com/.

National Institutes of Health. (2016) *Stem Cell Information.* Retrieved from www.nih.gov/.

Ogbogu, U. & Mengel, M. (2013) Who Owns Diagnostic Specimens in the Era of Personalized Medicine. *Canadian Journal of Pathology.* 5(3), 86–88.

Ogbogu, U. & Rugg-Gunn, P. (2008) The Legal Status of Novel Stem Cell Technologies in Canada. *Journal of International Biotechnology Law.* 5(5), 186–199.

Okita, K., Ichisaka, T. & Yamanaka, S. (2007) Generation of Germline-Competent Induced Pluripotent Stem Cells. *Nature.* 448(7151), 313–317.

Osborne, P. H. (2003) *The Law of Torts.* Toronto: Irwin Law.

Reference re Assisted Human Reproduction Act, 2010 SCC 61. Retrieved from https://scc-csc.lexum.com/scc-csc/en/nav.do.

Regina v. Secretary of State for Health ex parte Quintavalle (on behalf of Pro-Life Alliance). (2003) *UKHL 13.* Retrieved from www.parliament.uk/.

Rugg-Gunn, P. J., Ogbogu, U., Rossant, J. & Caulfield, T. (2009) The Challenge of Regulating Rapidly Changing Science: Stem Cell Legislation in Canada. *Cell Stem Cell.* 4(4), 285–288.

Samuel, G. (2005) Can the Common Law Be Mapped? *University of Toronto Law Journal.* 55(2), 271–297.

Schleifer, K.H. (2009) Classification of Bacteria and Archaea: Past, Present and Future. *Systematic and Applied Microbiology.* 32(8), 533–542.

Takahashi, K., Tanabe, K., Ohnuki, M., Narita, M., Ichisaka, T., Tomoda, K. & Yamanaka, S. (2007) Induction of Pluripotent Stem Cells from Adult Human Fibroblasts by Defined Factors. *Cell.* 131(5), 861–872.

Takhtajan, A. (1973) The Chemical Approach to Plant Classification with Special Reference to the Higher Taxa of Magnoliophyta. In G. Bendz (Ed.), *Chemistry in Botanical Classification* (pp. 17–28). Stockholm: Nobel Foundation.

Waddams, S. M. (2003) *Dimensions of Private Law: Categories and Concepts in Anglo-American Legal Reasoning.* Cambridge: Cambridge University Press.

Wade, N. (2013, October 14) The Clone Named Dolly. *New York Times.* Retrieved from www.nytimes.com.

Wertz, D. C. (2002) Embryo and Stem Cell Research in the United States: History and Politics. *Gene Therapy.* 9(11), 674–678.

Wilmut, I., Beaujean, N., De Sousa, P. A., Dinnyes, A., King, T. J., Paterson, L. A., Wells, D. N. & Young, L. E. (2002) Somatic Cell Nuclear Transfer. *Nature.* 419(6907), 583–587.

Xiao-Jie, L., Hui-Ying, X., Zun-Ping, K., Jin-Lian, C. & Li-Juan, J. (2015) CRISPR-Cas9: A New and Promising Player in Gene Therapy. *Journal of Medical Genetics.* 52(5), 1–8.

Zhang, Y., Lamb, B. M., Feldman, A. W., Zhou, A. X., Lavergne, T., Li, L. & Romesberg, F. E. (2017) A Semisynthetic Organism Engineered for the Stable Expansion of the Genetic Alphabet. *Proceedings of the National Academy of Sciences.* 114(6), 1317–1322.

7 Stem cells and nanomaterials as experimental kinds

Melinda Bonnie Fagan

Introduction

The concept of a 'stem cell' is a peculiar one, encompassing two very different ideas: that of a *cell*, and that of a *stem*. A cell is a well-understood biological kind – a living entity that can be picked out from its environment, with standard parts and clear boundaries. A stem, on the other hand, is the beginning of a growth process, the origin for something that is to be. The term 'stem cell,' at first glance, refers to both entity and process; a cell defined by what it gives rise to rather than its observable traits. This dual conception leads to challenges for classifying and individuating stem cells. But perhaps these challenges are a special case of a more general issue: challenges of classification in experimental or synthetic sciences. This chapter explores that possibility, by comparison of classification issues in stem cell biology and nanoscience.

In scientific practice, members of the kind 'stem cell' are conceptualized and individuated differently from members of the familiar biological kind 'cell.' The latter, traditionally considered the simplest 'unit of life,' exhibits strikingly clear individuation criteria. These individuation conditions are articulated by the tenets of Cell Theory:[1]

i   Cells reproduce by binary division.
ii An individual cell's existence begins with a cell division event and ends with either a second division event (producing two offspring) or cell death (and no offspring).[2]
iii Generations of cells linked by reproductive division form a lineage.

In practice, single cells are individuated by an enclosing membrane.[3] So individuation of cells is topological; the cell is the membrane and that which it bounds. This criterion determines, for any case of interest, how many cells are present. When cells reproduce by division, the process is finished when the membranes 'pinch off' to separate. So biologists have a fairly clear, unambiguous way of determining whether an entity is a member of the kind 'cell.' Indeed, cells are arguably our clearest example of a biological individual: a living entity that can be counted, picked out from its environment, and distinguished from other individuals of its kind. The kind 'cell' is relatively unproblematic; a well-behaved natural kind.

The kind 'cell' subdivides into many more finely individuated kinds, which robustly exhibit different clusters of molecular, biochemical, morphological, and functional characteristics. There are actually two distinct classificatory systems for cells: one for single-cell organisms (microbes), the other for cells making up the bodies of multicellular organisms.[4] In this chapter, I will be concerned with the latter only.[5] Cells comprising the 'building blocks' of multicellular organisms are classified into a number of *cell types*; for example, neurons, muscle, and blood cells. These major cell types are further subdivided into more finely grained types: the various types of neurons, blood cells, and so on. The number of cell types in an organism ranges from three to hundreds, depending on species and on how fine-grained a characterization is wanted. Any multicellular organism can, in principle, be decomposed into a heap of its constituent cells, which cluster into types.[6] So it is *prima facie* plausible to suppose that stem cells are one cell type among others. Just as for major cell types, there are many varieties of stem cell: embryonic, hematopoetic, neural, induced pluripotent, and many more. It appears that stem cells can be classified in the way of major cell types, subdividing the generic kind 'stem cell' into an array of more specific sub-types. If this were so, then stem cell classification would not present any special problem, and 'the stem cell' could be regarded as a well-behaved natural kind, like the more inclusive kind 'cell.'

However, as I have argued in previous work, stem cells are not conceptualized and individuated in the way of familiar cell types (Fagan, 2016a). Cell types are defined in terms of traits had by their individual members: cell morphology, biochemistry, metabolism, molecular surface markers, and gene expression profile. Stem cells, in contrast, are defined not in terms of traits they have, but by what they can transform into. More precisely, stem cells are defined as cells with (i) the ability to self-renew, or produce more stem cells by cell division; and (ii) the ability to differentiate, or to produce specialized cells representing later stages of development. For example, the most recent edition of *Essentials of Stem Cell Biology* emphasizes just these two abilities: 'Stem cells are functionally defined as having the capacity to self-renew and the ability to generate differentiated cells' (Melton, 2013, p. 7).[7] Similar definitions can be found in influential review articles and informational materials provided by major stem cell research organizations in Europe and the Americas: for example, Ramalho-Santos and Willenbring (2007), the Cell Therapy and Regenerative Medicine Glossary (2012), the website of the European Stem Cell Network (2016), the NIH Stem Cell information page (2016), and the website of the International Society for Stem Cell Research (2016). This general scientific definition invokes not traits had by individual cells, but reproductive and developmental processes that individual cells may engage in as part of a lineage.[8] So the concept of a stem cell is unlike that of the standard cell types, involving a cell lineage generated by processes of reproduction and development.

There are, as noted earlier, many more specific varieties of stem cell, sub-types of that general kind: adult, embryonic, epiblast, germline, hematopoietic, mesenchymal, neural, trophoblast, and many more.[9] To my knowledge, there has been as yet no attempt by stem cell biologists to systematically classify the various types of stem cell.[10] Indeed, a number of the types found in the voluminous stem cell literature overlap or crosscut one another, and some commentators have noted that the field's terminology is strikingly confusing and disordered (e.g., Rao, 2004). The broad clinical aims of stem cell research make identification and individuation of stem cells vital to the field's success. The overarching goal of stem cell biology today is to harness stem cell capacities for clinical applications, to ameliorate a wide array of pathological conditions (heart attack, blindness, cancer, neurodegenerative diseases, spinal cord injury, and diabetes – to name a few). In order to achieve any part of this goal, stem cell therapies must satisfy regulatory standards set by policy-makers.

Although these standards vary across political/regulatory contexts, one very common and well-motivated requirement is that stem cells for clinical use must be characterized so as to produce robust and predictable effects. As stem cells are defined in terms of their reproductive and developmental potential, their classification in these terms is a precondition for their clinical use. But the current method of doing so, without any common classification or nomenclature, makes for inefficient and uneven progress toward clinical goals. Rao (2004) notes two projects in particular that have been hampered by classificatory mayhem in stem cell biology: the search for 'stemness' genes, and settling the controversy over adult (tissue-specific) pluripotent stem cells. In the former, the diversity of cell populations identified as 'stem cells' leads to non-replicable results in genomic screens. As Rao puts the problem, 'It is hard to imagine what cells one would compare to identify "stemness" genes given the widely differing definitions of stem cells' (Rao, 2004, p. 453). In the absence of uniform classification, any search for genetic mechanisms underlying stem cell capacities, and thus for genomic targets of drug development, must begin from scratch for each identified stem cell population. The disparity in starting cell populations blocks cumulative and collaborative work across experimental set-ups. Controversy over the differentiation capacities of adult stem cells has exacerbated politicized debates about embryonic stem cell research, fomenting divides within the field and slowing progress in myriad ways. More generally, confusion about stem cell classification fragments the field, preventing researchers from gaining useful knowledge from their peers. A clear, simple, and uniform classification system for stem cell types would ease communication across subspecialties within stem cell research, and allow robust generalizations (if there are any such) to emerge from comparative studies.

In this chapter, I propose an abstract framework for classifying stem cells. Rather than imposing philosophical accounts of classification and natural kinds on this young science, this framework aims to make explicit classificatory notions implicit in the experimental practices of stem cell research. After presenting the framework, I illustrate its key points with several examples from stem cell research. I then compare stem cells so classified to nanomaterials, looking to Bursten's (2018) account of the latter. I conclude with some reflections as to whether these two categories of scientific entity, stem cells and nanomaterials, should be considered experimental or synthetic kinds.

A classification scheme for stem cells

In previous work, I have proposed a minimal abstract model clarifying the concept of a stem cell (Fagan, 2013a, 2013b, 2016a).[11] The minimal model is an explication of the general definition of 'stem cell,' constructed to be as simple as possible while capturing all currently known stem cell varieties. It represents processes of cell reproduction and development in terms of four variables:

i   cell lineage L, consisting of individual cells related by reproduction
ii   a set of variable characters C, with respect to which cells are compared across generations
iii a number of cell divisions n (alternatively, a time interval t)
iv   developmental process D, which orders cell states $s_1, \ldots, s_n$ from less to more specialized

I briefly discuss and motivate the inclusion of each variable, and then show how this model provides a simple, general classification scheme for stem cell varieties.[12]

Because stem cells are defined in terms of cell reproduction and development, the concept presupposes that of a cell lineage. A lineage, in general, is a complex entity extending over multiple generations of reproducing lower-level entities. A *cell* lineage, as noted earlier, is a biological entity composed of successive cell generations, organized by reproductive relations – that is, cell division. Self-renewal and differentiation processes presuppose not only a cell lineage, but also a system of comparisons across the generations comprising that lineage. Consider self-renewal, the conceptually simpler process. A cell with the ability to self-renew can divide to produce either one or two cells that resemble it, the parent.[13] Because no two numerically distinct cells are the same in absolutely every respect, a scientifically useful notion of comparison across cell generations must presume a set of variable characters relative to which such comparisons are made (e.g., size, shape, or expression level of a particular gene). For a stem cell, self-renewal means production of another stem cell that resembles its parent. So the characters to be compared across generations are just those taken to define the stem cell of interest. The very idea of self-renewal, then, involves a classification of cells within a lineage L as having the same values for some set of characters C. Variable n (or time interval t) specifies the number of cell reproductive events constituting the lineage.

Any lineage can be abstractly characterized as a tree-diagram with a particular number and arrangement of generations, branch-points, and end-points (termini). If stem cells were defined in terms of self-renewal alone, each would be characterized by a lineage L generated by cell reproduction over $n + 1$ generations, consisting of all and only stem cells resembling the original in characters C. But the concept of a stem is complicated by inclusion of a developmental process as well. In contrast to self-renewal (cell division with no change in characters of interest), differentiation entails change in some cell characters. But differentiation is not simply the converse of self-renewal. For one thing, the process of change in differentiation does not necessarily involve cell division. To fully explicate the concept of differentiation, we need to add the notion of a *cell state*. A cell state, briefly, is a pattern of gene expression and molecular interactions that determines a cell's structural and functional characteristics. Differentiation involves change in cell state, while self-renewal is cell division with no change in cell state (with respect to the characters of interest).[14]

Alongside change in cell state, differentiation involves at least two further aspects: (i) cells in a lineage *diverge from one another* in their traits, and (ii) these diverging changes in cell state are *directed toward* the specialized traits of a particular cell type, such as neuron, red blood cell, and so forth. We can think of these two aspects of differentiation as 'backward' and 'forward' looking, respectively: diversification from a common stem, and progress toward a developmental endpoint (or endpoints). In stem cell research, the forward-looking aspect tends to receive more emphasis; as a trait of stem cells, it is referred to as 'developmental potential' or 'potency.' So the concept of differentiation presupposes some idea of a developmental process, which provides a basis for ordering cell states as more or less specialized. The divergence of differentiating cells from one another corresponds to the branching pattern of a cell developmental lineage. For any such lineage, the origin-point is a stem cell, termini are different standard cell types, and the arrangement of branch-points indicates the developmental pathways that cells traverse in differentiation. The number of termini of these pathways is a measure of the initiating stem cell's developmental potential. (I return to the notion of tree model structure ahead.)

According to this model, a stem cell is a cell that is the origin of a cell lineage L, generated by n cell divisions and organized by comparison of characters C, which distinguish stages of a developmental process D of ordered cell states $s_1, \ldots s_n$ (Figure 7.1). Any cell to which values of these variables can be assigned such that this complex condition is satisfied is a cell that is capable of both self-renewal and

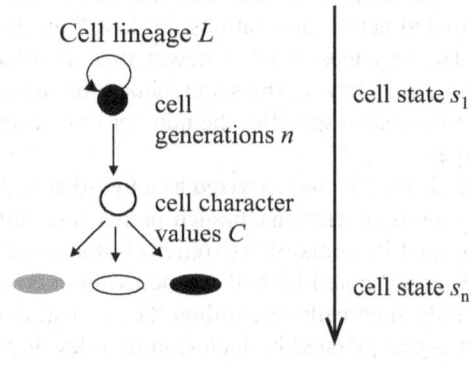

Figure 7.1 Diagram of the minimal stem cell model

differentiation, and thus a cell that conforms to the general definition of 'stem cell.' But stem cell researchers usually do not only want to know whether a given cell (or cell population) qualifies as a stem cell in the generic sense. They want to know exactly what its self-renewal and differentiation capacities are. Key questions asked of any proposed stem cell type are: How many other stem cells can it generate, and for how long? What range of cell types can the stem cell of interest give rise to? What are the characteristic traits of stem cells with these abilities; that is, how can we identify this type of stem cell in advance? Answering these questions amounts to discovering robust patterns among the values of variables L, C, n, and D. As a first step, these variables provide a classification schema for stem cell types, within which such patterns can be detected. A few concrete examples will illustrate.

Example 1: human embryonic stem cells (hESC)

hESC were first successfully cultured in 1998 by a multi-national group led by James Thomson at University of Wisconsin (Thomson et al., 1998). Their method set the standard for identifying cells as hESC, which with some modification is still used today. The method begins with a human blastocyst (a very early embryo) in artificial cell culture. Blastocysts have an outer layer (trophectoderm), which can give rise to extra-embryonic tissues, and an inner cell mass (ICM), which can give rise to embryonic tissues. This early embryo ICM is the source of, and specifies, lineage L. Next, a chunk of ICM is removed and placed onto a 'feeder layer' of cultured mammalian cells (first mouse, later human, and now replaceable by defined culture medium with a few chemical additives). In this new artificial environment, some of the ICM cells divide rapidly, producing clumps of 'outgrowth' after one or two weeks. Clumps are picked off the plates, dissociated, and replated, where new clumps appear in turn. Of these secondary colonies, a few with 'uniform undifferentiated morphology' are selected for further cycles of colony formation, self-renewal, and selection onto a new plate.[15] This continuous 'passaging' ensures that cells retained in the experiment are self-renewing; n (or t) is, therefore, as long as the experiment continues. In their new environment, some of the cultured cells divide to produce colonies with specific morphological and molecular characteristics, including rapid division in culture, lack of specialized traits, flat round shape, large nuclei surrounded by correlatively thin cytoplasm, prominent nucleoli, high telomerase activity, and high expression of particular genes (Thomson et al., 1998). A suite of molecular traits has been added to this profile in the subsequent two decades. This profile specifies the value for variable C. Further experiments showed that cells with these traits can, when placed in culture environments conducive to differentiation along particular pathways, produce more specialized cell types representing all the major germ layers. This experimentally detected differentiation potential specifies the value for variable D.

In summary, the details of Thomson et al.'s method and its results assign values to variables of the minimal stem cell model. Cells that are produced by that method and exhibit the specified characteristics and abilities are identified as hESC. Lineage L is specified in terms of species (human), developmental stage (early blastocyst), and location within the source organism (ICM). The traits exhibited by self-renewing cultured cells capable of differentiating into all the major cell types of the human body specify values for the set of cell characters C. The requirement for continued cell division entails that there is no upper limit for variable n; in practice, a threshold of fifty cell divisions was used to identify hESC. Differentiation potential is tested by moving some of these self-renewing cells to environments conducive to differentiation, and comparing results with characters of ordinary differentiated cells. These details of the method partly specify the values of variable D – the traits that distinguish more from less differentiated cell states.

Example 2: hematopoietic stem cells (HSC)

HSC are the blood-making, or hematopoietic, stem cells found in mammalian bone marrow (and, in lower concentrations, in circulating blood and several other organs). Many other 'tissue-specific' stem cells have been characterized as well, in brain, gut, skin, hair, eye, testes, and muscle. But the method for doing so was pioneered with HSC. In the early 1950s, it was discovered that lethally irradiated mice can survive lethal doses of gamma-radiation if injected with bone marrow cells from a 'donor' mouse shortly afterwards (see Kraft, 2009; Fagan, 2013a, Chapter 8). A bone marrow transplant effectively transplants the entire immune system, the cells of which are very radiation-sensitive. James Till and Ernest McCulloch of the Ontario Cancer Institute tried to develop a way to quantitatively measure that sensitivity, in mouse bone marrow cells. In the course of their research, they noticed that 'rescued' mice had nodules on their spleens (~1 million cells/nodule). When dissected, these splenic nodules were found to contain all known blood cell types – including those with immune function.[16] Till and McCulloch used chromosome markers to show that each nodule was a clone or colony descended from a single donor cell. They also showed that cells from spleen colonies could themselves rescue irradiated mice and produce new splenic colonies – that is, spleen colony cells could 'cycle' through multiple hosts (Till and McCulloch, 1961; Siminovitch et al., 1963). So, they reasoned, these colony-founding cells are the stem cells of the immune (and blood) system: hematopoietic stem cells. They repurposed their assay into an experimental system for detecting HSC: the 'spleen colony assay.'

Till and McCulloch found an approximately linear relationship between number of cells injected and number of spleen colonies formed (~1 to 10^4). So then the question became which of those 10^4 cells was the 'colony-forming unit,' that is, the stem cell?[17] This set off a wave of spleen colony experiments, aimed at finding just the right combination of characters to separate bone marrow cells into different sub-populations (or 'subsets'), looking for the pure population of blood stem cells. In the late 1980s, a group at Stanford University presented a modification of Till and McCulloch's method using single-cell sorting technology coupled with in vitro blood cell culture assays (Spangrude et al., 1988; discussed in Fagan, 2013a, Chapter 8). The resulting characterization of HSC has proved fairly robust, although molecular refinements have been added in the succeeding decades. The basic procedure is as follows. First, bone marrow cells are collected from an adult organism (assigning the value for variable L) and sorted into sub-populations according to size, density, surface molecules, and cell cycle status. Next, these sub-populations are classified according to values of morphological, functional, and molecular characters, including cell size, density, and presence or absence of surface molecules.[18] The sorting procedure assigns distinct values for variable C to each 'cell subset.' Differentiation potential is then tested by moving samples of each subset into an environment conducive to blood cell differentiation along a particular pathway. The details of these experiments specify values for variable D. Because of the decisive role of whole animal survival, the time interval of interest (variable t) for HSC experiments extends at least six months in mice, and in humans, decades. HSC are defined as the subset containing all and only the bone marrow cells that can give rise to all the major blood cell types and reconstitute the immune system.

Example 3: induced pluripotent stem cells (iPSC)

iPSC is a generic stem cell kind, encompassing many sub-types that are produced by variations on the same general method (reviewed in Maherali and Hochedlinger, 2008). That method is 'direct cell reprogramming,' pioneered by Shinya Yamanaka's research team at Kyoto University (Takahashi and Yamanaka, 2006). These experiments begin with putting differentiated cells in artificial culture. Their differentiated origin assigns a value to variable L. A few (two to four) specific gene sequences are added to these cells' nuclei, after which a very small percentage (~0.05%) transform to resemble embryonic stem cells of the same species.[19] For example, differentiated skin cells form a flat, evenly distributed layer of fibers in artificial culture. But after these genes are added, altering gene expression patterns, a few of the skin cells transform to exhibit a different morphology: round and clumped, like embryonic stem cells. Additional molecular and biochemical traits are also measured (cell surface molecules, activity and expression of specific proteins, expression of specific genes, global gene expression, and histone modification of specific genes). These measurements, collectively, assign a value to variable C. These transformed cells are maintained as self-renewing colonies, just as for embryonic stem cells (ESC); this sets variable n as the duration of the cell line. Differentiation potential for iPSC is also established in the same way as ESC, fixing the value of variable D. The transformed cells capable of producing all the major types of cell making up the mammalian body are identified as iPSC (of a particular subtype, reflecting the precise details of the cells' molecular profile and developmental capacities).

A similar account could be given of all the known stem cell varieties. In this way, the minimal model provides a schema for classifying all the various kinds of stem cell: adult, embryonic, pluripotent, induced, neural, epiblast, and so on. These different kinds correspond to different assignments of values of the variables L, n, C, and D (Table 7.1). Although the details that assign stem cells to various types differ, experiments that characterize stem cells have a basic structure

Table 7.1 Model-based classification of stem cells

Type	L (source)	C	n	D
ESC	5d embryo ICM	cell size, cell shape, gene expression, karyotype, telomerase activity, alk-phos, cell surface molecules	≥ 50 divs	traits of cells from three germ layers
HSC	BM, cord, peripheral blood	cell size, density, light scatter, surface molecules, cell cycle status	> 6 months	traits of main blood and immune cell lineages
NSC	basal lamina of ventricular zone	cell morphology, surface markers, gene expression, cytokine response	months to years	traits of neurons, astrocytes, and oligodendrocytes
iPSC	various (relatively mature cells)	colony shape, cell size, cell shape, nucleus/cytoplasm ratio, cell surface molecules, activity and expression of specific proteins, gene expression (specific and global), histone modifications at key locations	≥ 50 divs	traits of cells from three germ layers
GSC	5–9 wk gonadal ridge	colony shape, alk-phos, surface expression (SSEA-1, SSEA-3, SSEA-4, TRA-1–60, TRA-1–81)	20–25 wks	traits of cells from three germ layers
EC	teratocarcinoma (129)	cell shape, morphology, production of embryoid bodies, surface molecules, enzymes	unlimited	traits of cells from three germ layers, teratocarcinoma

of three stages (Figure 7.2). In the first, cells are extracted from an organismal source (specifying *L*) and placed in a new environment in which candidate stem cell characters *C* are measured. The duration of this part of the experiment specifies the value of *n* (time or cycles of self-renewal). In the second stage, measured cells are moved to another environment in which capacities for differentiation can be realized. Finally, characters of differentiated cells (the termini of *D*) are measured. So details of the experimental procedure fix values of *L*, *n*, *C* and (partly) *D*. Experiments with this basic design are the only way stem cells are identified, in actual practice.

The details of different experimental materials and methods specify different values for variables in the model. Not only are values of variables comprising this classification schema assigned by experiment, but also the features corresponding to the variables themselves are parts of the basic experimental method. This classification scheme differs in an important way from that associated with familiar differentiated cell types, such as 'neuron.' Crucially, the features used to distinguish among kinds of stem cell include features of experimental methods and their results, not just traits had by the cells themselves. We have no other access to stem cells than these experiments. In this sense, stem cells are inherently

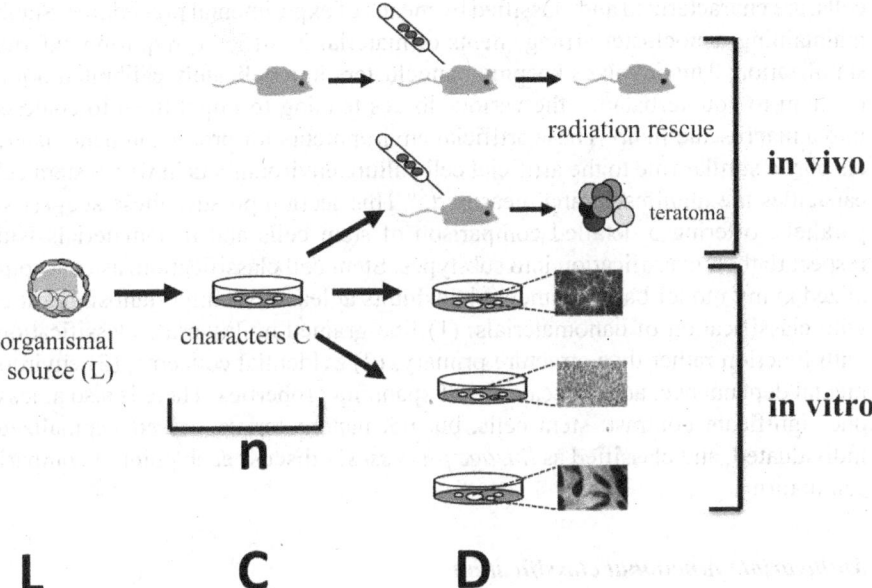

Figure 7.2 Schematic structure of stem cell experiments, showing specification of variables of the minimal model

experimental entities, and the various types of stem cell can be considered experimental rather than natural kinds.

Stem cells and nanomaterials: comparison

I next compare stem cells, as understood in terms of the classification scheme presented earlier, to nanomaterials, as characterized in recent work by Julia R. S. Bursten (Bursten, 2016, 2018; Bursten et al., 2016). Nanomaterials are composed of unit particles at the nanoscale, this being typically defined as approximately 1–100 nanometers (10^{-9} meter) in at least one dimension. Such nanoparticles are composed of a few hundred atoms of a particular chemical element or metal oxide (e.g., gold, carbon, titanium dioxide), arranged in a cluster. Strikingly, these cluster assemblies exhibit physical properties distinct from macroscale assemblages of the same element, and (less surprisingly) from individual atoms of that element. These scale-dependent properties include conductivity, catalytic activity, ductility, and color. At the nanoscale, then, elemental composition alone does not determine structure or function. This has obvious implications for classification of nanomaterials, which accordingly differs from traditional chemical classification.

The contrast suggests some parallels with stem cell classification, which similarly differs from traditional classification of cell types. The parallelism is further supported by the 'synthetic' character of nanoscience. Nanomaterials, like stem cells, are characterized and classified by means of experimental procedures. Stably maintaining nanocluster arrangements of material is difficult, requiring external stabilization. This involves keeping nanoclusters in a delicately calibrated liquid medium to counterbalance the various forces tending to impel them to coalesce into a macroscale lump. These artificial environments for producing nanomaterials play a similar role to the artificial cell culture environments in which stem cell capacities are manifested and measured.[20] This section pursues these suggestive parallels, offering a detailed comparison of stem cells and nanomaterials with respect to their classification into sub-types. Stem cell classification, as conceptualized in my model-based framework, exhibits at least four important similarities with classification of nanomaterials: (1) fine-grained multivariate classification, with function rather than structure primary; (2) evidential concerns; (3) environmental dependence; and (4) scale-/level-spanning properties. There is also at least one significant contrast: stem cells, but not nanomaterials, are conceptualized, individuated, and classified as *lineage processes*. I discuss each point of comparison in turn.

Multivariate functional classification

As discussed earlier, stem cells are defined in terms of cell reproduction and development, conceptualized as functional capacities attributed to cells: self-renewal and differentiation potential. Classification of cells exhibiting these capacities into one of the stem cell types requires specification of the values of at least four variables: lineage source L, cell characteristics C, number of cell divisions n (or time interval t), and developmental process D. Nanomaterials as a class are also defined in functional terms; that is, reactive chemical or physical properties such as catalysis and conductivity. Moreover, classification of nanomaterials also requires specifying the values of four variables (each of which may include more fine-grained information): chemical composition, size, shape/dimensionality, and surface chemistry (Bursten, 2018; see also Chapter 8).

Composition, in terms of constituent chemical elements and their molecular arrangement, is obviously important – but, on its own, insufficient to specify the behaviors that are central to nanomaterial classification. Dimensionality refers to the number of dimensions of a material that extend beyond the nanoscale (100nm); nanoclusters are 0-dimensional, long nanotubes or wires 1-dimensional, and thin nanofilms 2-dimensional. The behavioral properties of a nanomaterial vary widely according to dimensionality. Size within the nanoscale also makes a difference, with small nanoclusters of > 5nm exhibiting behavior and patterns of molecular geometry than larger (50–60nm), more crystalline spheres of the same composition. Surface chemistry also plays a key role, balancing instability at the nanomaterial's boundary with external agents that maintain its shape and size over time. So kinds of both stem cells and of nanomaterials are distinguished by a tetrad of variables; a classification scheme that includes fine-grained information about the entities in question: lineage, cell characters, cell divisions/

time, and developmental process; composition, dimensionality, size, and surface chemistry. However, the two fields differ in the way this fine-grained classification relates to the functional properties that define the more generic kind ('stem cell' or 'nanomaterial'). For stem cells, the fine-grained classification specifies in more detail the very functional capacities that define the generic kind 'stem cell' (alongside cell and organismal characteristics; see ahead). However, the four-part classification of nanomaterials does not include the functional properties that mark something as a nanomaterial. Rather, the tetrad of variables is taken to determine those properties. In this way, nanomaterials science conforms to what Bursten refers to as 'the structure – function paradigm' in chemistry, while stem cell biology does not.

Evidential concerns

Another similarity between stem cell and nanomaterial classification is that both schemas raise evidential concerns, due to the combination of 'destructive characterization' and heterogeneity in the relevant experiments. In previous work, I have referred to these concerns in the field of stem cell research under the heading 'stem cell uncertainty' (Fagan, 2013b, 2015, 2016b). Briefly, the problem is that it is impossible to experimentally demonstrate that one and same individual cell is a stem cell; that is, that it possesses both the capacity for self-renewal (for some value of n) and the capacity for differentiation (for some value of D). This is because, in order to show that a cell is capable of self-renewal, an experimenter places it in a special environment that inhibits differentiation, and observes whether that cell divides to make more stem cells. And to show that a cell is capable of differentiating into a particular kind of differentiated cell, an experimenter places it in another environment, which supports differentiation into that cell type ('along that pathway'). It is not possible to perform both experiments on a single individual cell.

That negative conclusion follows from single-cell individuation conditions and the two generic defining stem cell capacities (see earlier). Self-renewal means that the starting cell has been replaced by its progeny. Differentiation means that a cell is no longer a stem cell (and also, usually, it has been replaced by progeny). So stem cell experiments involve a form of 'destructive characterization' – the cell about what the classification question is asked ceases to exist during the course of the experiment. To show that a given cell is a stem cell, and to classify it as one type or another, both capacities need to be realized. But this dual experiment cannot be performed on a single cell. So stem cells are perforce identified and classified at the population level. But this does not obviate the evidential concern, because stem cell populations are notoriously heterogeneous and prone to unpredictable fluctuations in phenotype.[21] A population of identical stem cells (a clone) allows experimenters to effectively test 'the same cell' across a range of different environments, severally conducive to self-renewal and the differentiation pathways under consideration. But this means that what we identify as a stem cell and classify as a particular type thereof is also relative to the assumption that the population of cells we are testing is homogeneous; that they are of a kind with no internal variation relevant to the capacities of self-renewal and differentiation. Although this assumption is unavoidable, it is in practice often found to be false.

One reason for this is that there is no pre-determined list of phenotypic characters that pick out all and only stem cells, of any variety; experimental researchers must work this out for themselves. They do so via repeated use of the basic experimental method described earlier, using different combinations of cell surface markers (sets of closely related values for *C*), so as to gradually reduce mismatch with stem cell capacities. Stem cell researchers try to specify more and more finely the population of all and only stem cells of a particular type – to keep adding markers to get closer and closer to that pure population of, say, blood-forming stem cells in inbred mice. So hypotheses about stem cells, even when confirmed by experiment, are always provisional. As new cell traits are discovered and made accessible to measurement, the assumption that the cell population under test is homogeneous must be continually reassessed, and is often revised. Another factor that mitigates against the homogeneity assumption is that stem cells are highly *plastic* – that is, they are prone to alterations of phenotype, for reasons that are not currently well-understood. These plastic tendencies sharpen the evidential concern raised by the evidential constraint on single-cell classification noted earlier.

Although the details of course differ, classification of nanomaterials faces similar problems.[22] Destructive characterization is a feature of certain experimental methods used to characterize a given nanomaterial, particularly techniques aimed at revealing composition and crystallographic structure. Characterization techniques such as mass spectrometry, inductively coupled plasma atomic emission spectroscopy, and small-angle neutron scattering all aim to identify features of nanomaterial composition or surface structure. In each of these methods, information about these characteristics is obtained via analysis of induced emissions from the nanomaterial under investigation. But, in the process, the material makeup of the remaining sample is changed. Just as in stem cell experiments, the object of inquiry ceases to be that which is characterized, over the course of an experiment designed to characterize it.

In colloid synthesis methods, heterogeneity of outcomes of a single synthesis procedure is referred to as 'the problem of monodispersity.' This problem presents an additional evidential challenge for nanomaterials classification. Briefly, these methods use carefully calibrated solutions to allow a small number of atoms of a metal such as gold to nucleate into small nanoclusters, and then stabilize these clusters to prevent further nucleation. But nanoclusters of the same element, prepared in the same way (even within the same synthesis batch) can exhibit different molecular geometries, and thus different chemical and physical properties. Nanomaterials' tendency to heterogeneity has led some (including Bursten) to give up the idea of their having fixed essential characteristics at all. In any case, it is clear that experimentally based classification of stem cells and of nanomaterials face similar evidential challenges.

Environmental dependence

Another similarity is that classifying characteristics are (partly) dependent on features of the entity's environment. For stem cell capacities, environmental dependence is obvious. The previous section argues that stem cells are individuated and classified only relative to a particular experimental context.[23] In the course of any such experiment, stem cell capacities are revealed by altering the physical, chemical, and cellular environment of cells. To briefly recap: all stem cells derive from an organismal source that fixes the value of lineage L. Extracted cells are then placed in an environment conducive to self-renewal (n, t) and/or cell phenotype characterization (C). Last, differentiation potential is revealed by placing cells in one or more other environments, designed to encourage development along particular pathways (D). So the basic method of identifying and characterizing stem cells consists of moving cells from one environment to another. This method takes advantage of one of stem cells' signature characteristics: their traits and capacities – that is, their very identity as a stem cell – change according to features of the surrounding environment. Because of this environmental dependence, there is no 'absolute' or generally individuated stem cell type, as there is for, say, neurons or red blood cells. To put the point particularly trenchantly, stem cells are *not cells* – that is, they aren't individual 'units of life' that can be distinguished from their environment by a bounding membrane. They are instead the starting points of lineages associated with particular experimental contexts, in which multiple cell environments are implicated. The reproductive and developmental trajectory of any particular lineage depends largely (although not completely) on the environmental context of participating cells.

Relatedly, the source organism for a given stem cell, although not part of the experimental method per se, is also an environment for stem cells or their precursors. Every stem cell is a part of, or descended via cell division from a part of, exactly one multicellular organism. For the purpose of stem cell classification, that organismal source is characterized by species, developmental stage, and location of cell extraction.[24] Scientific names for different varieties of stem cell often refer to these features of the source organism; for example, human embryonic stem cell, mouse epiblast stem cell, and so on. Stem cells found within a post-natal organism (the so-called adult stem cells) are named for the part of that organism they build and/or replenish: neural stem cells, hematopoietic stem cells, epithelial stem cells, muscle stem cells, and so forth. More broadly, stem cell research has long been divided into two branches, which focus on 'adult' or 'embryonic' stem cells, respectively. The adult/embryonic distinction refers to the developmental stage of stem cells' organismal source.

Furthermore, the termini of the lineage a stem cell initiates are (or can become) parts of a multicellular organism.[25] In this way, the lineage generated by a given stem cell is embedded in, or flanked by, contexts of organismal development. In recent work, I argue that stem cell developmental potential involves not only a range of cell types, but also diverse modes of developmental organization in multicellular organisms (Fagan, 2017, 2018). At least six different modes of development (normal, pathological, minimal, dispersed, organ-like, and embryo-like), approximating different aspects of organismal development and organization, can be initiated by stem cells. Stem cells, therefore, are capable of not only differentiating into a range of cell types, but also realizing different processes or modes of differentiation. I refer to this capacity as 'developmental versatility': the ability to produce different aspects of organization associated with different modes of development. Which mode (or combination thereof) is realized in any particular developmental process initiated by a stem cell, depends on features of that cell's environment, and of its descendants' environment (e.g., presence or absence of specific biochemical signals, spatial/geometric arrangement of cells, physical boundaries of the system, and physical factors such as oxygen levels or pH). The general conclusion is that stem cell identity and its developmental consequences are environment-dependent. Stem cells differentiate in diverse ways depending on their environment.

Nanomaterials synthesized using colloidal methods (see earlier) show similar dependence on the environment for their structure and behaviors. As noted earlier, surface chemistry is an important factor determining the characteristic behaviors and specific structure of nanomaterials. Solutions of particles of a particular element (or metal oxide) are designed to allow nucleation to form nanoclusters, but not further agglomeration. Maintaining clusters of material in solution at the nanoscale requires stabilizing agents (termed 'capping agents') to counteract the various forces that would otherwise produce macroscopic lumps of material, with proportionately fewer atoms localized to the unstable compound surface. Bursten uses the metaphor of sausage casings, which 'hold together the atoms of a nanomaterial and maintain its shape over time' (Bursten, 2018, p. 15). The choice of capping agent, and the procedural details of how solutions are mixed over time and in precise proportions to produce nanomaterials, influence the size, shape, and structure of the resulting cluster – and, accordingly, its chemical and physical behavior. Solubility of nanomaterials is similarly environment-dependent (see Chapter 9). So although environmental-dependence is not as central to nanomaterial synthesis as it is to stem cell characterization (where the method itself is premised on this form of plasticity), some of the key variables for classifying nanomaterials involve dependence on specific features of the chemical and physical environment of developing nanoclusters. Environmental dependence is a feature of both classification schemes.

Level/scale transitions

A fourth similarity is the significance of 'level-' or 'scale-spanning' characters in both stem cell and nanomaterial classification. The scale-dependence of nanomaterial categorization is a major theme in Bursten's work on the subject (Bursten, 2016, 2018; Bursten et al., 2016; see also Chapter 8). Her key point, I take it, is that attending to classification of materials at the nanoscale offers a richer and more nuanced conceptual account of the relation between macroscopic and microscopic domains in chemistry. Important conceptual insights about classification across these domains are achieved by considering 'movement between scales;' that is, examining how chemical phenomena and our explanations thereof change when the scale of chemical entities under consideration shrinks or expands. Briefly, and following recent work of Wilson, Bursten argues that changing scale is a 'driving force for change in conceptual structure' (Bursten, 2018, p. 18). The ordered hierarchy of scales as it figures in nanoscience has at least a rough parallel in the various 'levels' of biological organization implicated in the stem cell concept: molecular, cellular, organismal. The notion of 'cell state,' implicated in developmental processes, posits that cell traits and behavior are determined by a complex, underlying molecular network. The basic idea is that a particular configuration of a molecular network of genes, RNA, protein, small molecules and biochemical groups, is what makes a cell the kind of cell that it is.[26] 'Cell identity' is often equated with cell state. Exercise of stem cell developmental capacities involves transformation of cell state. So it is no wonder that the very idea of a stem cell challenges traditional notions of cell classification.

Stem cell biologists often, in their more theoretically inclined moments, represent the process of development as a landscape (see Fagan, 2012). This way of thinking about development goes back to C. H. Waddington in the mid-20th century, who famously depicted the developmental potential of a tissue as an inclined landscape of branching valleys (Waddington, 1957). The inter-level relation of molecules and cells is visualized by the landscape's two sides: a top-side view of branching pathways leading to stable developmental states, and an underside of genes and their interacting products. We can think about cell development as the dynamics of these states. A cell starts in the stem state and from there enters one or another pathway, which leads to a differentiated terminus. The challenge going forward is to integrate tissue, organ, and organismal levels of organization into this emerging developmental model. This challenge, and its implications for classification of developing entities, seems akin to that of scale-transformations linking nanoscience with other domains of chemistry.

Contrast: stem cells as lineage processes

One important feature of stem cell classification, which does not seem have a parallel in nanomaterials, is that the multiple classifying features correspond to a *lineage process*. Lineages in general are modeled using 'tree diagrams' that track relations between generations of reproducing entities. An assignment of values to these variables corresponds to a space of possible lineage tree models. Each model within this space has a particular number and arrangement of developmental stages, branch-points, and termini – a particular structure, or 'tree topology.' The simplest such model is a linear sequence of cell states ordered in time. However, because cell developmental processes occur alongside (and may be coordinated with) cell division events, stem cell lineages tend to have the form of branching hierarchies. The number of distinct cell states determines the depth of the lineage hierarchy. The arrangement of branch-points indicates the arrangement of distinct developmental pathways initiated by a stem cell. The number of termini corresponds to a stem cell's developmental potential. So any given stem cell (or population of stem cells) can be characterized in terms of the structure of the corresponding tree topology the cell lineage that stem cell initiates:

- d, number of stages (depth of developmental hierarchy)
- p, number of termini (a cell's developmental potential)
- a, arrangement of branch-points (the number of which = p-1)

In stem cell experiments, d is fixed by the cell characters that are measured at different time intervals during the experiment; a by what we know about developmental pathways in the system in question (often very little); and p by the tests used to establish developmental potential.[27]

The lineage structure framework is just an alternative interpretation of the four-variable model presented earlier. As discussed earlier, nanomaterials too are classified in terms of a tetrad of features. But what is distinctive about stem cell classification in terms of lineages is that the latter concept is itself a powerful classification scheme for biology. Phylogenetic lineages are the prevailing, although not the only, mode of classification for organisms and species. Classification of the major cell types, as noted earlier, does not traditionally make use of lineages – but classification within those types often does (e.g., neurons, blood, and immune cells). Stem cells' distinguishing developmental capacity, the potential to differentiate into a range of other cell types (and, arguably, forms of organismal organization), can itself be deployed as a classificatory principle for cells of the associated lineage. In this way, stem cell classification 'loops back' to inform classification of more traditional cell types, and perhaps of developmental modes pertaining to tissue, organ, and organismal levels of biological organization. This secondary, or derived, classificatory function does not appear to be paralleled in the case of nanomaterials – for the simple reason that the latter do not generate more entities like themselves; they do not reproduce and develop, as living things do. Stem cells, by definition, are the starting points of cell lineages. The same features by which these peculiar entities are classified as varieties of stem cell determine a lineage structure than can, in turn, by used to classify cells within that lineage.

Conclusion: synthetic or experimental kinds?

I will conclude with some reflections on the implications of the earlier comparison for the notion of 'synthetic science.' In her (2018) article, Bursten argues that fine-grained classification involving multiple factors (variables) is 'endemic to nanoscience as a synthetic science,' and that classification in synthetic sciences departs from the clean-cut, theoretically articulated taxonomies traditionally assumed by philosophers of science. Instead of these traditional taxonomies, nanoscientists try out different combinations of values for the four features used to classify nanomaterials (composition, size, dimension, and surface chemistry), in order to synthesize new nanomaterials with desired properties. The fine-grained classification scheme for nanomaterials, combined with the 'structure-function paradigm,' serves as a fruitful heuristic guide for making new materials. Given nanoscientists' synthetic purposes, 'greater specificity of structure is well worth the cost of a clean-cut taxonomic framework' (Bursten, 2018, p. 21).

Although, as noted earlier, stem cell research does not at this time proceed under an analog of chemistry's structure-function paradigm, researchers' emphasis on connecting multiple levels of biological organization (molecular, cellular, organismal) suggests that the field is working toward such a framework. Furthermore, Bursten's description of nanoscience as a synthetic science also applies to important areas of stem cell research, such as cell reprograming (see Example 3 in this chapter). The method of producing induced pluripotent stem cells is applied with subtle variations in organismal source, cell culture environments, and genetic reprograming factors, to yield a wide array of different stem cell lines, calibrated for different purposes. Stem cell science as a whole, I have argued, involves classifying stem cells according to a fine-grained schema of four variables (lineage, cell characters, cell divisions/time, developmental process), assignment of values to which determines a cell lineage with a particular structure and content. Should we then say that stem cell biology – the whole, not just reprogramming – is a synthetic science? And if so, why is the field not allied with or included in the emerging field of synthetic biology?[28]

The similarities between classification in stem cell biology and nanomaterials science offer reason to characterize both as 'synthetic.' But there is at least one countervailing reason; that is, reason to think that stem cell biology is not, properly speaking, synthetic. This is not to say that stem cell biology is a science in the way philosophy of biology has traditionally considered; it evidently is not. But rather than 'synthetic,' the better term to describe the field as a whole may be 'experimental.' The countervailing reason has to do with the aims of stem cell biology, and with the role of multicellular organisms as 'flanking' stem cell lineages. As noted earlier, every stem cell derives from an organismal source, features of which determine a stem cell's lineage *L*. It is true that, in the course of classifying any stem cell as such, that cell (more precisely, cells from the population identified as that kind of stem cell) is manipulated by transfer to multiple artificial environments, which produce both self-renewing stem cells and a range of differentiated cells. The results of these experiments can fairly be called 'synthetic.' But these results are not the ultimate goal of stem cell research. That goal is, instead, to provide replacement parts for human bodies or to coax our bodies into making those parts for ourselves. The results of stem cell research are intended for use in the body – they are intended to become *us*. Although this ultimate goal remains elusive (apart from a few special cases, such as bone marrow transplantation therapy), it remains central to, and arguably even constitutive of, stem cell biology as a distinctive scientific field.[29]

Because stem cells are derived from multicellular organisms and intended to return to them, there is an important sense in which these entities are 'natural,' albeit experimentally manipulated. This does not obviate the important similarities discussed earlier, between stem cell and nanomaterial classification. But it raises the question of whether the features Bursten highlights as characteristic of synthetic science, are better understood as features of new, experiment-driven scientific fields. What, if anything, separates classification in *synthetic* sciences from classification in *experimental* sciences? Both have been neglected by traditional philosophy of science. On this point, as for so many other questions, our philosophical classification of kinds of scientific inquiry may be refined, by closer examination of actual scientific practices.

Notes

1 See Schwann (1847) for an early statement of cell theory. The tenets presented here are updated to reflect our current scientific context.
2 In sexually reproducing organisms, gamete cells can also fuse to form a new cell: the zygote. As stem cell phenomena occur within organismal generations, I discuss only mitotic divisions here.
3 Of course, signals from the environment pass through the cell membrane all the time, and vice versa – the membrane is not a barrier.
4 It is possible for one and the same entity to belong to both classificatory systems. See, for instance, recent discussions of 'the holobiont' and the immune criterion of selfhood (Pradeu, 2012).
5 See O'Malley (2014), and references therein, for more on microbial classification.
6 Cells are not the only constituents of multicellular organisms, nor is cell-level decomposition the only possible or scientifically significant breakdown of these organisms.
7 Melton goes on to expand on the general definition: 'A more complete functional definition of a stem cell includes a description of its replication capacity and potency . . . a working definition of a stem cell line is a clonal, self-renewing cell population that is multipotent and thus can generate several differentiated cell types' (Melton, 2013, pp. 7–9). This more elaborate functional definition retains self-renewal and differentiation as the defining capacities of stem cells.
8 The term 'stem cell' (rather, its German cognate Stamzelle) was introduced by Ernst Haeckel, who used it to refer to the unique cellular 'stem' of a multicellular organism – that is, the fertilized egg (Haeckel, 1905 [1877], I, pp. 130–131; Dröscher, 2014). Although the fertilized egg is not considered a stem cell today, the notion of cellular origin of parts of a multicellular organism, represented in the form of a lineage or tree, remains part of the current stem cell concept.
9 The index of *Essentials of Stem Cell Biology*, an influential textbook, references the following sub-types of stem cell: adult, amniotic/amniotic fluid, bone marrow, cancer, cardiac, cord blood hematopoietic, dental pulp, embryonic (human, primate, mouse), embryonic germ, embryonic kidney, embryonal carcinoma, epiblast, epidermal, (hair) follicle, germline, hematopoietic, induced pluripotent, intestinal, keratinocyte, leukemic, liver, mesenchymal, multipotent, multipotent adult progenitor, myogenic, neural, pancreatic, pancreatic liver, pluripotent, post-natal, renal, skeletal, skeletal muscle, solid tumor cancer, somatic, tongue, trophoblast, very small embryonic-like. This list is actually an underestimate of the number of distinct stem cell varieties circa 2009, as I have left out the different species designations and 'progenitor' cell types – the latter are sometimes considered stem cells, sometimes not, depending on experimental context.
10 There are attempts to classify certain broad classes of stem cell, for example, pluripotent or induced pluripotent (Kurtz et al., 2018), but not the entire range of stem cells.

11 The current version is somewhat updated from the original, but the basic ideas are the same. The main contrast is the explication of differentiation. The earlier treatment stipulated that each mature cell type is distinguished by a specialized set of character values M, which includes morphological, functional, and molecular characters. Cells in lineage L *differentiate* over some time interval t_1-to-t_2 just in case cell character values for M at t_2 are more similar to those of mature cells than at t_1. The present account is a further generalization of that earlier treatment.

12 For further background, see Fagan (2013a), Chapter 2.

13 Cell division events are classified as 'symmetric' if two offspring cells resemble each other; 'asymmetric' if they differ. Self-renewing divisions can be symmetric (both offspring resemble the parent) or asymmetric (one offspring resembles the parent).

14 Importantly, a cell state, as such, is not dependent on cell context – although which state a cell is in at a given time is thought to be largely determined by context, the state itself can be characterized without reference to the extracellular environment. Whether this concept is realistic is not fully established.

15 The original hESC protocol included a layer of cultured 'feeder cells' that supported the ICM-derived cells, both physically and by producing growth factors. Feeders can now be replaced by defined chemical media (2i).

16 Three kinds of blood cells were known at the time: myelo-erythroid, B and T lymphocytes.

17 The spleen filters the blood, so most of the injected cells would presumably pass through the spleen, although the 'retention rate' was unknown.

18 For more detail about these methods, see Fagan (2007).

19 Characters selected: colony shape, cell size, cell shape, and nucleus/cytoplasm ratio. The first reprograming experiments used four genes (Oct3/4, Sox2, Klf4, and c-Myc) and mouse skin cells.

20 Nanomaterials are prima facie more akin to stem cells produced in artificial culture than to those extracted from the bodies of multicellular organisms.

21 These variations and shifts are attributed to factors in individual stem cells' immediate microenvironment, or niche (see main text).

22 Thanks to Julia Bursten for information on destructive characterization techniques in nanoscience.

23 Indeed, classification of stem cells is doubly relative: to details of the experimental method specifying values of classificatory variables, and to the (provisional, defeasible) assumption that the cell population being tested in such an experiment is homogeneous.

24 Variable L can thus be broken down into sub-variables concerning the organismal source: species, developmental stage, and location of cell extraction within the organism. For example, for hESC: human, early (5-day) embryo, inner cell mass. Stem cell research could in principle investigate the full range of values of these three sub-variables, characterizing the properties of stem cells derived from different species, developmental stages, and organismal parts. But the field in practice is less systematic, and more oriented toward medical applications. Most stem cell research is done on just two species: mouse and human.

25 These can be the same organism, but in most cases, they are distinct.

26 Another way to put the point: the molecular cell state explains and individuates both traditional cell types, and the experimentally relative types of stem cell.

27 One way in which stem cell lineage trees differ from phylogenetic lineage trees, however, is that transitions between cell states are, at least in some cases, reversible. So another variable to be considered in characterizing a given stem cell in terms of the lineage framework is degree of reversibility (R).

28 See Endy (2005) and O'Malley (2009) for overviews and aims of synthetic biology.

29 Fagan (2013a, Chapter 10) argues this point in detail.

References

Bursten, J. (2016) Nano on Reflection. *Nature Nanotechnology*. 11(10), 828.

———. (2018) Smaller Than a Breadbox: Scale and Natural Kinds. *British Journal for the Philosophy of Science*. 69(1), 1–23.

Bursten, J., Hartmann, M. & Millstone, J. (2016) Conceptual Analysis for Nanoscience. *The Journal of Physical Chemistry Letters*. 7(10), 1917–1918.

Cell Therapy and Regenerative Medicine Glossary. (2012) Stem Cell. *Regenerative Medicine*. 7, S1–S124.

Dröscher, A. (2014) Images of Cell Trees, Cell Lines, and Cell Fates: The Legacy of Ernst Haeckel and August Weismann in Stem Cell Research. *History and Philosophy of the Life Sciences*. 36, 157–186.

Endy, D. (2005) Foundations for Engineering Biology. *Nature*. 438, 449–453.

European Stem Cell Network. (2016) *Stem Cell Glossary* [online]. Retrieved from www.eurostemcell.org/stem-cell-glossary#letters 13 March 2016.

Fagan, M. B. (2007) The Search for the Hematopoietic Stem Cell: Social Interaction and Epistemic Success in Immunology. *Studies in History and Philosophy of Biological and Biomedical Sciences*. 38, 217–237.

———. (2012) Waddington Redux: Models and Explanation in Stem Cell and Systems Biology. *Biology and Philosophy*. 27, 179–213.

———. (2013a) *Philosophy of Stem Cell Biology*. London: Palgrave Macmillan.

———. (2013b) The Stem Cell Uncertainty Principle. *Philosophy of Science*. 80, 945–957.

———. (2015) Crucial Stem Cell Experiments? Stem Cells, Uncertainty, and Single-Cell Experiments. *Theoria*. 30, 183–205. (Special Section: Philosophy of Experiment).

———. (2016a) Cell and Body: Individuals in Stem Cell Biology. In A. Guay & T. Pradeu (Eds.), *Individuals across the Sciences* (pp. 122–143). Oxford: Oxford University Press.

———. (2016b) Generative Models: Human Embryonic Stem Cells and Multiple Modeling Relations. *Studies in History and Philosophy of Science Part A*. 56, 122–134.

———. (2017) Stem Cell Lineages: Between Cell and Organism. *Philosophy and Theory in Biology*. 9. (Special Issue: *The Ontologies of Living Beings*).

———. (2018) Individuality, Organisms, and Cell Differentiation. In O. Bueno, R.-L. Chen & M. B. Fagan (Eds.), *Individuation across Experimental and Theoretical Sciences*. Oxford: Oxford University Press.

Haeckel, E. (1905) The Evolution of Man: A Popular Exposition of the Principal Points of Human Ontogeny and Phylogeny, Translated from the fifth (enlarged) Joseph McCabe (Ed.), 2 vols. (New York: G. P. Putnam's Sons, 1905) Original: Anthropogenie oder Entwicklungsgeschichte des Menschen. Gemeinverständliche wissenschaftliche Vorträge über die Grundzüge der menschlichen Keimes- und Stammes-Geschichte, (Leipzig, 3rd edition 1877),

International Society for Stem Cell Research. (2016) *Stem Cell Glossary* [online]. Retrieved from www.isscr.org/visitor-types/public/stem-cell-glossary#stem 8 June 2017.

Kraft, A. (2009) Manhattan Transfer: Lethal Radiation, Bone Marrow Transplantation, and the Birth of Stem Cell Biology, 1942–1961. *Historical Studies in the Natural Sciences*. 39, 171–218.

Kurtz, A., Seltmann, S., Bairoch, A., Bittner, M. S., Bruce, K., Capes-Davis, A., Clarke, L., Crook, J. M., Daheron, L., Dewender, J. & Faulconbridge, A. (2018) A Standard Nomenclature for Referencing and Authentication of Pluripotent Stem Cells. *Stem Cell Reports*. 10, 1–6.

Maherali, N. & Hochedlinger, K. (2008) Guidelines and Techniques for the Generation of Induced Pluripotent Stem Cells. *Cell Stem Cell*. 3, 595–605.

Melton, D. (2013) Stemness: Definitions, Criteria, and Standards. In R. Lanza & A. Atala (Eds.), *Essentials of Stem Cell Biology*. 3rd edition (pp. 7–17). San Diego: Academic Press.

National Institutes of Health, U. S. Department of Health and Human Services. (2016) Glossary. *Stem Cell Information* [online]. Retrieved from http://stemcells.nih.gov/glossary/Pages/Default.aspx 13 March 2016.

O'Malley, M. (2009) Making Knowledge in Synthetic Biology: Design Meets Kludge. *Biological Theory*. 4, 378–389.

———. (2014) *Philosophy of Microbiology*. Cambridge: Cambridge University Press.

Pradeu, T. (2012) *The Immune Self*. Oxford: Oxford University Press.

Ramalho-Santos, M. & Willenbring, H. (2007) On the Origin of the Term "Stem Cell". *Cell Stem Cell*. 1, 35–38.

Rao, M. (2004) Stem Sense: A Proposal for the Classification of Stem Cells. *Stem Cells and Development*. 13, 452–455.

Schwann, T. H. (1847) *Microscopical Researches into the Accordance in the Structure and Growth of Animals and Plants*. London: The Sydenham Society.

Siminovitch, L., McCulloch, E. & Till, J. (1963) The Distribution of Colony-Forming Cells among Spleen Colonies. *Journal of Cellular and Comparative Physiology*. 62, 327–336.

Spangrude, G., Heimfeld, S. & Weissman, I. L. (1988) Purification and Characterization of Mouse Hematopoietic Stem Cells. *Science*. 241, 58–62.

Takahashi, S. & Yamanaka, S. (2006) Induction of Pluripotent Stem Cells from Mouse Embryonic and Adult Fibroblast Cultures by Defined Factors. *Cell*. 126, 663–676.

Thomson, J., Itskovitz-Eldor, J., Shapiro, S., Waknitz, M., Swiergel, J., Marshall, V. & Jones, J. (1998) Embryonic Stem Cell Lines Derived from Human Blastocysts. *Science*. 282, 1145–1147.

Till, J. E. & McCulloch, E. A. (1961) A Direct Measurement of the Radiation Sensitivity of Normal Mouse Bone Marrow Cells. *Radiation Research*. 14, 213–222.

Waddington, C. H. (1957) *The Strategy of the Genes*. London: Taylor & Francis.

Dialogue four

Synthetic kinds in chemistry and nanoscience

8 Nanochemistry meets philosophy of science

A conversation about collaborative classification

Jill E. Millstone and Julia R. S. Bursten

Introduction

The following is an interview of Jill E. Millstone, a nanochemist, by Julia R. S. Bursten, a philosopher of science. Millstone and Bursten collaborated actively from 2011 to 2015, during which time Bursten served as resident philosopher in Millstone's nanosynthesis laboratory, while Millstone served as a member of Bursten's dissertation committee (Bursten, 2015). This collaboration established a mutual interest in the principles underlying the classification of nanomaterials, which has influenced both of their research interests and inspired projects in both of their research programs (see, e.g., Bursten, 2016 Bursten et al., 2016; Bursten, 2018).

The interview that follows is lightly edited for readability. It addresses a number of topics that the researchers have found critical to their projects in nanomaterials classification, and it centers on the four-part scheme that Millstone uses to classify nanomaterials. This scheme identifies the composition, size, shape, and surface chemistry of a nanoparticle and, in so doing, specifies the physical and chemical properties expected of the particle. This scheme has been defended to philosophical audiences in Bursten (2018). The interview also makes extensive use of analogies and disanalogies between nanochemistry and other areas of chemistry and engineering, particularly organic chemistry and materials science, as well as offering an illustrative example of collaboration between a philosopher of science and a nanochemist.

Establishing a collaboration

BURSTEN: Let's warm up a little bit by talking what it was it like for you to have a philosophy graduate student contact you. I wrote you asking for permission to take a class you were teaching, and you asked me to come meet you before allowing me into the class.

MILLSTONE: When you first contacted me for the class, I did not know if this was going to work, because I wasn't sure if you had any science background. I was concerned that you might not be able to do the class, but then when you were talking about your background and how you had a lot of physical science training I thought, "OK, well, let's see. That should work." Then, when we started talking, I just immediately was really fascinated with the kinds of questions that you were asking because it was obvious to me that conceptual analysis was such a big part of what would help us think about what we're doing.

For instance, we try to challenge ourselves by saying, "What is a mechanism in nanoscience?" My students would frequently want to understand a particular mechanism, but what does understanding a mechanism even look like? In organic chemistry, we have a pretty good idea of what a mechanism looks like: descriptions of electron pushing, rates, and so forth. But even there, why does that represent the complete understanding of the system? *Is* that the complete understanding the system? I was pretty excited about the things that you were talking about.

You were obviously extremely knowledgeable too. I don't want to leave out that component. You had a strong physical sciences background but also, you know philosophy really well, I mean, every question I asked, you had an answer for. It wasn't like, "Oh, let me get back to you." We had a fluid interaction for the entire time we met, and all the times I asked you questions, so that was a big deal. But once I realized you were an intellectual force, that got me really excited about the things we could do.

BURSTEN: Did you know anything about philosophy of science was before we started talking?

MILLSTONE: Zero percent. At that point, I didn't even know it was a discipline.

BURSTEN: Are there one or two things that stick out to you as the most significant or important things that have come out of learning about philosophy of science?

MILLSTONE: Asking the kinds of questions you ask, like not only "What is the evidence?" but "Why is this evidence?" I don't know if I could give you one singular thing. Certainly, it's been so useful in explaining to people why studying nanoparticle surfaces is challenging and interesting. The conceptual concerns about surfaces give much more to me, and are a much more comprehensive and complete explanation of why studying surfaces is challenging, than just saying it's tough analytically. Wolfgang Pauli famously said, "God made the bulk; surfaces were invented by the devil," and I find that so eye-rolling, first because it is a useless explanation of the difference between the two, and second because it takes something – the notion of a surface – that is so impactful in our world, and so fascinating from the chemistry perspective, and makes it sound bad because it's hard. Importantly, I think part of what is hard is that people don't know how to talk about surface concepts and don't know how to think about it. And that's why I really feel like you need philosophy of science to really talk productively about surfaces. But none of that is bad.

Practically speaking, for me and my group, the things that stand out are you asking things like, "Why does that convince you?" "What is it about that evidence that seems convincing?" and also asking, "What do you mean by this or that?" "What does that mean?" You keep pushing and pushing to try to understand what a concept means. That forces us to be much more scientifically clear, and it usually pushes us to the point that we don't know an answer, but we know what we don't know and what we need to know, and that gives us the idea for an experiment. It's a very practical series of questions that way. So I think just pushing on what we're saying a lot is a different way than you typically do chemistry, and it's very helpful.

BURSTEN: It's funny, you've just described the Socratic method. "But what do you really mean? What do you really mean?" That's what he did.

Collaborative philosophy of science

BURSTEN: Can you say in your own words what it looks like to talk about surfaces with philosophy of science? What is different about the way we talk about it together and the way that I talk about it than the Pauli picture?

MILLSTONE: That's a really good question because it has a really clear answer. So there's two different ways to talk about surfaces in chemistry. We think about how much energy it takes to create a surface per unit area, so we talk about surface energy. We talk about curvature and changes to surface energy as a function of curvature. We talk about reactivity as a function of curvature. These are all looking at a surface from a continuous descriptive type of model, so there we're not thinking about the individual atoms. And then we can also think about the individual atoms, such as thinking about how many unpassivated bonds are there on an atom at the surface of a material.

It becomes clear to most people familiar with the chemical sciences that those two ideas about a surface, they can be bridged sometimes. For example, you can describe surface energy by calculating the number of dangling bonds per unit area, the energy associated with each bond, and get some estimate. When we use these terms of "surface area" versus "dangling bond," the different terms give us different suites of tools that we are going to use to describe that system. You and I have talked about this point a lot. For an Au_{25} cluster, people routinely talk about surface atoms, but for a cholesterol, which has just as many (or more) atoms, we would never talk about its surface. For the nanoparticle, because it's gold, because the core is metal atoms, the ideas of surface chemistry are immediately in a chemist's mind. I think what philosophy of science does is give me the way of talking about why we are applying a particular model, what does that model get us, when we talk about surfaces with respect to Au_{25}. That was language that you used to use in group meeting: what does applying this particular model get us in terms of predictive power, and what does it not get us in terms of predictive power? Just pointing out that the two models envision the system differently is something that we didn't used to think about on a day-to-day basis.

We can recognize this type of disconnect in so many different types of terminology that we use in chemistry. We've applied a term indiscriminately through a series of systems and never asked, "How is it scaling?" This is quintessentially what you philosophers of science are studying, and this is a huge thing you've brought to us. Most of the time when you ask those kinds of questions and I look at it, the term actually isn't doing the work I thought it was doing, or it's applying a set of conditions to my system that may or may not be true, and I haven't asked whether they're true or not. It might be true, it might be OK, but I haven't asked the question, which I need to. So I think with surfaces, it's a good example of where I'm applying these models just out of habit. Now I'm asking, "What have I left out of the system when I apply a particular model, and what does that mean about the model?" When we've asked those questions to our scientific collaborators, we've gotten a very good response from people. I don't think that kind of question is something that we ask in science routinely, at least in chemistry.

Conceptual challenges in nanosynthesis

BURSTEN: I know you think there are some things about nano that make it exceptionally amenable to this kind of question. What are some of those things? Why does it work to ask these questions about nano? Do you think asking these questions in an organic chemistry setting would go over similarly, or do you think that there's something about the interaction between this kind of analysis and the subject matter?

MILLSTONE: Why it wouldn't work as well in organic chemistry settings is what we were just talking about. In organic chemistry, the models give them the right answers, and so they don't change models. There's no reason to think "should I be thinking about the surface chemistry of a cholesterol?" because their observations and syntheses are well-described by the theories they use right now. So if they did start thinking about the surface energy of a molecule, it probably wouldn't bring new capabilities for them. I think asking these questions wouldn't be as impactful as it is for nano, or for materials chemistry in general. Although, maybe it would, and that's all the more reason we need philosophers of science helping us to think about these things.

In materials, you have this idea that the atomic structure is responsible for the macroscopic properties –

BURSTEN: The structure-property paradigm.

MILLSTONE: Exactly. That's so ingrained in our understanding of materials and the way that we approach making materials that it's almost hard to think about the fact that it isn't necessarily a straight path. Even for me who thinks about it a lot, I'm almost uncomfortable talking about it too much. Contemplating the idea that there might not be a direct relation between structure and properties almost feels like sand shifting. In nanoscience, you're so often straddling between molecular chemistry and materials science, reading whole journals on crystal growth and then going back and looking at coordination complex reactivity. You're so often straddling these different ways of describing how things form and what synthesis looks like that it's almost a relief to have language to talk about how you have to combine these two worlds and that they don't always combine seamlessly. I think nano being at that disciplinary boundary makes this series of questions resonate with us, but it also makes the philosophical lens particularly impactful for addressing the scientific questions that we have.

BURSTEN: The way I tend to think about organic synthesis is as a sort of assembly of known parts and a guess-and-check procedure. The model says the parts will behave a certain way if I apply energies at certain appropriate intervals. Then if my model says it's going to go one way and the experiment fails to go that way that means, either I've applied the model wrong or I've done something wrong in the synthesis. There are well-established relationships between the synthetic protocol, the model, and the iterative guess-and-check work of the process of experiment. Nano is not like that.

MILLSTONE: There's a lot complexity in organic chemistry reactions, so combining the known parts is an intellectual challenge. It is difficult to combine the parts. But the physical and organic chemistry principles that guide organic synthesis are largely well-established. You find new instances of where they apply or don't apply, but for the most part, you're not fundamentally changing how we understand a carbon-carbon bond. You're trying to find more and more creative ways to exploit and augment the suite of models that describe organic molecules, and to do more and more advanced things with them. That's very exciting and hugely impactful.

A chemist considers classification

BURSTEN: Before we go into nano, can you say a little bit about the role that classification plays in that kind of intellectual work you just described in organic chemistry?

MILLSTONE: There are so many shorthands in trying to predict an organic chemistry reaction: what is the electron-withdrawing character of this group? How localized are the electrons? What is the nucleophilicity of this species? And so on. There are all kinds of markers by which I would evaluate a particular molecular backbone to see where and how I could construct or build from it. So every time, I'm taking my starting material, and I have a goal for my final material. I can name every reagent that goes into my synthesis and I can name every product that comes out. It may be that every product that comes out is stuff I don't want or didn't expect, but every product that comes out is identifiable and nameable.

There's an organic molecule naming convention.[1] Everything is named, which means that you know what your target is, and you know when you hit it and when you don't. You also know what every starting material is. Relative to nano, it's very straightforward to trace why I didn't get a product that I wanted and explain what happened because in organic chemistry, I can definitively identify everything in the reaction.

Now, an organic chemist reading this might object that you can't identify every intermediate, and that's technically true. However, that the reason you can't is because sometimes they don't exist for long enough for you to isolate them and study them using the analytical methods available. It's not that if you isolated them, you still wouldn't know what it means. That's the big difference. Organic chemists can have trouble identifying their intermediates, but that's a tool problem, not a chemistry problem.

The state of nanochemistry right now is analogous to the state of organic chemistry in 1895–1900. Organic chemists in the early 20th century could already make complex molecules like artificial indigo and aspirin. With the complexity of those molecules, you could easily see people being like, we already know what all we need to know, why do we need to keep studying organic chemistry? It's a comment I hear all the time about nano: we can already make all kinds of nanomaterials, why do we need to keep studying them? To me, that's a little bit of a lack of vision. Just because you can make complex architectures doesn't mean that you have any ability to control those architectures, to then drive the chemistry to where you need it to go.

For example, take progestin. It is a transformational organic molecule, and relatively complex: it's a fused-ring structure with several chiral centers. It's also the main component of birth control, which has effects way, way beyond what we do in the lab. It's a culture-shifting molecule. That kind of synthetic capability didn't just spring up after we made indigo. It took 60 years of hard work to understand things like how the electronegativity of atoms influences the reactivity of a bond, implications of functional groups on neighboring atom reactivity, and how to determine and exploit reaction kinetics, among so many other major chemical discoveries both in the organic and physical chemistry spaces. Through all those innovations, you also had huge innovations in analytical spectroscopies, nuclear magnetic resonance imaging, mass spectrometry. All these different techniques had to come online, and that all fed into, and also fed from, the development of organic chemistry. So synthetic studies impact way more than just the target. But because people put in that work, and they made the effort to understand the physical chemistry that underlies organic reactions, now natural product synthesis is hugely advanced. We can make a huge, huge variety of molecules. Yes, there are still challenges, with very, very smart people working on them. Organic chemists are trying to create molecules that could be cancer-saving based on architectures in a hermit crab that took 3.5 billion years to evolve. That's where you want science to go. You want to be able to say, "I want this molecule and then I have the tools to go into the lab and make it."

Right now, we don't have this in nano. We cannot say, "I want this nanoparticle and these are the tools I have to make it." What I mean is that we don't have the ability to say what nanoparticle we want to make, which sometimes means we don't even have the ability to clearly individuate or identify what "this nanoparticle" even refers to, let alone to say what synthetic route will make it. We have some rough ideas: there are two classes of nanoparticles that are showing promise for various reasons, semiconductor nanoparticles and metal nanoparticles. Within those classifications, we know of the existing particle species, which perform well in their canonical applications. We can say what particles might perform a little bit better in those applications, but to me, that's a just a small dream.

It's a small dream because it's just taking a few steps from where we already are. We have a whole periodic table; we have a whole universe of materials. What do those elements and materials look like when they're synthesized at the nanoscale? What do they look like when I control every aspect of the particles? We are not anywhere close to doing that. It could be revolutionary. Of course, it could also be useless, but I have a really hard time believing that to be true.

This brings us right to the point of what you're talking about: classification. What part of the nanoparticle do I have to control? What part of it is impactful? That changes as a function of the size, shape, and surface chemistry of the particle. Synthetic control is a moving target. What part of the nanoparticle we have to define, what part we have to control, is not the same for every nanoparticle. Right now, we don't know what nanoparticles need what to be controlled, and of course we also need to know *for* what, controlled for what? The challenge is that it's so, so, so, complex. That's why it's super exciting to me. If I had to move in a narrower space, I think I would be really sad.

These aren't problems that can be tackled by one person, but the key thing is, I don't think a lot of people are asking these questions. I don't say that as a self-aggrandizing thing. I think it is lack of interfacing with people outside the lab. It's a lack of interacting with the bigger picture questions here. We are so application-driven in the modern scientific community. In nano, a common grant evaluation is whether something is useful for a battery. If not, or if not right now, move on. Is it useful for a solar cell? No, moving on. So there's not a lot of space to ask these bigger questions, and there's also not a lot of interactions with people where these questions would come up. If we could get behind asking these questions as a community, I think it would be a way to really make use of this class of materials, which is undoubtedly housing a huge number of technological breakthroughs. We may never get there if we don't step out, because these things have to be answered. I don't know how you make something if you don't know what you're trying to make.

BURSTEN: Well, yes and no. People made metal nanoparticles by melting finely divided metals into molten glass for years without ever knowing what they were. They used them to an effect, making color in glass with localized surface plasmons, that we can now isolate and control. I mean obviously you know I'm on board with this vision but there's an interesting way in which –

MILLSTONE: But isn't that a case in point though? Until we knew what it was, we couldn't really leverage it. We could make stained glass windows, but it wasn't until we knew what plasmons were that we could study their properties and determine the glass color was a plasmon. Then we though, "Oh, we can use that for cancer therapy," "Oh, that's useful for photocatalysis," you know.

Classification in nanosynthesis

BURSTEN: OK, yeah, good point. So nanochemists are controlling and tuning the properties by understanding how to generate them. All the parts, what the parts are, are moving and shifting in nano. Can you give an example of one synthesis or one material where you see this shifting happening?

MILLSTONE: Of course. My wheelhouse is gold. I've worked with gold most of my life. We work with a lot of different materials now, but gold is still the home example.

BURSTEN: The gold standard, if you will.

MILLSTONE: Ha, sure, the gold standard. So, for gold, let's think about gold particles that have between 150 and 200 atoms. I can start asking about how a ligand on the surface of that gold particle changes the total electronic structure of the particle. If I look at the electronic structure – and here I'm talking about the density of states – over the entire particle what I find is that most of the changes are occurring at the surface. This is not a surprise, because of the ligands. How does that then impact, for example, its catalysis or its photophysical properties, such as its luminescence.

Now, let's say I have a gold nanoparticle that has 200 atoms and it's passivated with ligand A, so there's 100 ligand As around it. Suppose I take one of those gold atoms and I replace it with a copper. How does that influence the properties that I observed, the photoluminescence that I observed from the particle? If I took cholesterol and I replaced one of the carbon atoms with phosphorus, it would completely change the function of cholesterol in the body. Organic chemistry is riddled with these very small changes, thalidomide obviously being the most one of the most famous ones. So we know that will change the function completely. Now, I'm going to make a change of a similar magnitude on this 200-atom particle. I'm going to change one of the gold atoms to a copper. I haven't told you where I'm going to change it yet. Let's say I change it at the dead center. What's going to happen? Calculations indicate that not much is going to change.

However, let's say I do it at the surface, so that one of the ligands now is bound to copper. Then I start to change the observed optical and electronic properties that I got from the particle. At that length scale of 200 atoms, I can see this difference of copper, but only when I put the atom in a particular place. It's not as exaggerated as it is if I had Au_{144}, but it is still present. Whereas if I go up even to 300 gold atoms and replace one, I've basically lost the effect. Take it again: instead of a gold particle, now I have a copper particle. I do the reverse: I replace one copper atom with gold. The effects can be completely different. We don't synthetically have the control to do this comparison experiment, but if you do it in on the computer, you see big changes.

There's a couple of things I want to highlight from this example. First, the substitution of an atom in a nanoparticle has an impact that scales with the size of the nanoparticle. Second, the magnitude of the impact of the position of that heteroatom also scales as a function of the size of the particle, but it will also correlate with the shape of the particle. If I had been talking about a triangular plate as opposed to a sphere, the impact of its substituting a surface atom would have changed from substituting it on a sphere. Third, in all of what I just said, I've ignored the fact that these ligands are binding to a surface. Their binding motif impacts where the copper can go from an energetic perspective, and it also changes the motif of the ligand. So as the radius of curvature changes, regardless of the electronic structure, that changes the ligand motif and so on. I actually have to write these things out to keep track of all the different things I'm trying to talk about.

So the copper atom replacing one atom, the impact of replacing one atom in a nanoparticle versus a molecule, the impact of the position of that atom and the complexities of the position of the atom is going to make a difference in whether we observe an impact on that electronic properties or not. The magnitude of that impact is dependent not only on size but also on shape, and then we had this whole other variable of surface chemistry.

The relationships between surface chemistry and these other variables are hard to articulate; they are what I talk to you about. I think part of that difficulty is because there's so much we don't know. It is both driven by and drives particle shape, and it is both driven by and drives heteroatom incorporation, and so the balance between those drivers changes for a particle too. And, that's all looking at a particle in isolation. Now let me add on the fact that every nanoparticle is a kinetic product, so the pathway by which I access it, the synthetic route, also determines a lot about whether or not I can access any of these architectures, which architectures are trapped, which are formed, and so forth. So it's a separate question from the structure that we were talking about, but it's closely related. That also brings up the fact that for nanoparticles, they are kinetic products that can rearrange in time.

BURSTEN: Right, relaxation time is huge. Or, rather, it's very short, which is huge. This is great, but the funny thing is that as you start to articulate all of these different moving pieces it starts to sound more and more like organic chemistry.

MILLSTONE: Does it?

BURSTEN: Yeah, it starts to sound more and more like we know all of the pieces and it's just a matter of sort of putting the pieces together, if the pieces are things like size, shape, surface chemistry, and heteroatom presence.

MILLSTONE: Oh, OK, right, so the key thing to say is that we just know that those things are factors. We don't know how they factor in.

BURSTEN: So we don't have a story for nano, like we do in organic chemistry, of chemical understanding of a thing like electron distribution. That's the thing that drives a lot of organic mechanisms, and we have it down pretty well, and yeah there's kinetics as well, and that understanding is there too. With organic chemistry, we know what all the different moving pieces are about and how they relate to each other, which is exactly what we don't know in nano, and what we're trying to figure out. And the story in organic chemistry is that, one way or another, it's all about electron motion. That's the story for general-chemistry reaction theory as well, and also in a different way for inorganic, like, large inorganic complexes are all about electron motion. Do you think Nano is going to be all about electron motion?

MILLSTONE: No, it can't be. Nanoparticles, whether we're talking about semiconductor nanoparticles or metal nanoparticles, the particulate part of that, that
is a phase change. It is not a chemical reaction, so it can't just be about that.

BURSTEN: I love that distinction. Can you say more?

MILLSTONE: When you're making a nanoparticle in lab using wet chemical techniques, everything looks like traditional chemistry. You have a flask, you
have a Schlenk line, all of that kind of stuff. You're reducing a metal or you're
doing some kind of decomposition reaction. You're doing chemical transformations during that step, but how those reduced metal cations, or those
decomposed precursors, then turn into a solid phase suspended in a liquid
medium, a colloid: that's a phase change. We don't know how those reduced
metal ions go from their reduced state into a nanoparticle. The best models
that we have right now talk about that as a phase change. It is a solidification
process, nucleation. Nucleation is a phase change. So it fundamentally can't
be all about electron pushing, following an electron around, because that's
not what makes a nanoparticle.

BURSTEN: I hadn't thought about it in quite those terms before. That's so useful.

MILLSTONE: Oh good.

Synthesis, computational modeling, and scientific infrastructure

BURSTEN: One of the other things that came up for me while we were talking was
to look at the way nanoparticles are modeled on computers. Because of the
time in history that we're in right now, the relationship between modeling
nanoparticles on computers and synthetic abilities in the lab is very different
from the relationship between either physical or mathematical modeling of
large organic molecules, and synthetic processing in the lab, that they had in,
say, the 1950s.

MILLSTONE: Well, we are more advanced than the 1950s were. 1950s computational ability to organic chemistry is not modern computational ability to
nanochemistry, but we still don't have the computational power we would
need to model nanoparticles atom by atom. Because nanomaterials include
so many more atoms than a typical organic molecule, and if you include the
ligands, which you really need to, then there is a limit to electronic structure
methods. We can model things by molecular dynamics, but it's not really
molecular dynamics that we're often interested in.

This might seem like it conflicts a little bit with what we were just saying, but
let me clarify why it doesn't. For the final particle, we want to know about its
electronic structure. We need to include as many components of that particle as
possible, especially on the smaller end of the nanometer length scale, things with
at least one dimension around 5 nm or below. Molecular Dynamics can easily
handle thousands of atoms, but the phenomena it can tell us about here are things
like metal segregation. If we knew anything about the connection between the

chemical change steps and the physical change steps maybe we could do some work modeling nanoparticle formation reactions this way. The problem is we don't have a lot of inputs right now for that model. I forgot where you were going with that, but anyway I feel like we have some good density-functional methods for computer modeling of smaller nanoparticles, but it's really difficult to get past around 100 to 200 atoms with that. That's about as high as you can get, which is hard because that's not very high.

BURSTEN: Do you think that there's anything epistemically different about the sets of relations between models of chemical structure, synthetic protocols, and imaging techniques that we have now for nano, relative to a similar time in the history of other chemistry disciplines? Like you were saying, the 1895–1900 organic, or the sort of 1950s height of organic structure discovery. It seems like these three sets of relationships really drive the way that we think about what our next steps are for theoretical and synthetic advancement.

MILLSTONE: I hate to sound pessimistic here, but I don't think we have a lot of room in modern science for doing the fundamental physical chemistry that would be necessary to really understand what's happening in a nanoparticle reaction, to get that theoretical and synthetic advancement. There's a lot less nano going on today than there was even two years ago. There are still people who are earnestly trying to figure out how a nanoparticle forms, but it's really few, and it's becoming fewer every day. There's really no enthusiasm in funding it. Even in the scientific community, editors treat "nano" as a useless buzzword – partly rightly after years of abuse. That attitude combined with the fact that nanomaterials synthesis, for the vast majority of its 25-year history, has been driven by application, application, application, there is just almost no one who is asking about how nanoparticles form in the way that we are asking it and talking about it.

My research is largely about fundamental understanding. The physical and inorganic chemistry that goes into converting molecular precursors into a solid phase, what does that look like? That requires an enormous analytical effort to understand. I don't see much enthusiasm for that. There's this idea that nano had its time and it didn't produce anything and so the chemical community is moving on. I genuinely think it's just because we live in a different time.

In the 1950s, 1960s, and 1970s, there was Bell labs. This was a company-sponsored playground for scientists, and huge things came out of it. No one was telling these guys, "Find a transistor, find a transistor, find a new way to communicate." It was nowhere on the map, because truly impactful science is very hard to anticipate. If it is iterative, if it follows from all the steps that I just took, that is not transformational. The kind of physical chemistry work and the type of heyday at that time, I don't know if nano is ever going to get to go through that because it does require resources. It needs time, it needs staff, and it needs analytical capabilities.

I was reading forward-looking work this past spring from a group of scientists who would widely be considered some of the most expert nano-people in the country. Nowhere in the work was the word "nano." Literally, it was nowhere. They were doing absolute linguistic gymnastics to avoid saying the prefix "nano-." It was eye-popping. They want to do the work, but they don't want to say the word. They think other scientists and funding agencies see nano as passé.

The other thing is that I don't see is a lot of enthusiasm for finding out the fundamentals. I see people trying to take a class of materials, and say that they will look at the band gap of all the materials in this combinatorial array of syntheses, and they're going to screen for a particular property, like in the pharmaceutical industry. That's a good way to find things that match your application target. It's not a good way of understanding the chemistry of how you made that target. So I see people moving in the direction of trying to do large-scale applications-targeted syntheses, but no step of retroactively trying to understand what that might tell us about how nanomaterials form. I just genuinely don't think people see (or feel that there is) the opportunity in general.

I'm not trying to talk about myself as a visionary; I feel like a lone crazy person being like, "No, no, it's really great!" but I think part of it comes from this idea that we don't think enough about how unique nano is in terms of how it takes and blends existing knowledge, and how that also means that there is very likely a whole field of knowledge that we don't know. To me, it is exactly like space exploration. I know where large objects in space are, yes, and I have a vessel by which I'm traveling through space. That doesn't mean I understand the universe. I feel like the opportunities are huge, but, we'll see. Maybe you and I will figure out the language to use to help people get over these pseudo-vision leader statements about where things are going and what things are passé.

Defining "nano"

BURSTEN: I want to ask some more chemistry questions, but we're on this little linguistic point – well, it's not little, actually, it's really important in terms of the social structure of science. As someone who works in nano, how bad for you was it that Apple chose to name an iPod "nano"?

MILLSTONE: That's a really interesting question. For me, it was neutral or slightly positive because it least it brought attention to nanotechnology. Where is the nano in that? It's in the transistors. What does that mean? What does it take to get you your iPhone XS? The whole electronics industry is really one of the few areas of innovation at the nanoscale where you can easily point to both the incredible innovations that have happened, and where we are hitting the limit. At 3 nm, you're talking about tens of atoms. Transistors are running up against physical limits, there are so few atoms in there. A top journal editor can poo-poo nano, but the guys at Intel, they're thinking long and hard about what physics is changing when you go from 3 nm to 2 nm. That gives me some heart, because practically it's super important to get a handle on nanoscale material behavior, if only for electronics. And when we see how important it is in electronics, I think the optimistic view would be that we will take note and see how it will be impactful manipulating nanoscale phenomena in other material systems.

Even if you don't think about nanoparticles, this length scale is tremendously important. If you think about engineering plastics, or you think about engineering composites, artificial bone, artificial teeth, artificial anything, the length scale on which nature is doing its most important activities, it's all on the nanometer length scale. Whatever you want to call it, whatever level of interest you want to have in it, that doesn't change the universe. It doesn't change what will be impactful. It doesn't change where the opportunities are. Being able to tailor a material with these dimensions will be impactful. But did Apple confuse the public about what nano was? Yes.

BURSTEN: I was thinking more about, do you think the fact that "nano" became a branding tool for things that had nothing to do with nanotechnology contributed to the attitude that led the editors at top journals to dismiss it as a buzzword?

MILLSTONE: Oh, yes, absolutely, 100%.

BURSTEN: I think your answer was actually more robust than the question I was asking.

MILLSTONE: No, but it's from the National Nanotechnology Initiative, in 2000. Wherever you put money, you're going to create an incentive to abuse a term. So everybody in science then is going to try to call their stuff "nano" because they want to get in on this initiative, because that's where the dollars are. But are you not going to feed the hungry because there might be somebody standing in line just taking advantage of it? I don't think it's a good idea to forget it and close shop just because it was profitable to say you were doing nano and people took advantage. We should have check steps against this, though, and we can learn from nano here. It should be in the peer-review process or in the grant review process. If you're talking about a 500-nm polystyrene bead that doesn't have any size-, shape-, or surface chemistry-dependent properties, if we can't really classify within the group of size-dependent materials, you shouldn't call that nano. There wasn't a lot of self-policing.

BURSTEN: But without classification structure . . .

MILLSTONE: Exactly. That is the thing and you could almost just end there. Without knowing what nano is, it can become so diffuse that it undermines the science that we might otherwise have.

BURSTEN: I think there's a sense in which nano is a bit like pornography, in that you know it when you see it.

MILLSTONE: Yes.

Classifying nanomaterials: composition, shape, size, surface chemistry

BURSTEN: That's great because it encourages some of the really productive interdisciplinary projects that we were talking about earlier, and the ability to access a variety of different models. Like when you're talking about surfaces, if you have people from different communities thinking about the challenge of modeling surfaces differently, it gets more creative approaches than saying the problem of surfaces is a problem that is only going to be solved by a certain strategy or community. Without something to call it, without the ability to call it the surface of something, the idea gets so diffuse that it's impossible to have a conversation because we don't know what the associated groups of properties and models are. It's the opposite of the paradigmatic chemical classification example of the periodic table.

MILLSTONE: Yeah, I mean, the periodic table is miraculous.

BURSTEN: Right, it has this highly organized set of structures that all relate to each other in reasonably systematic ways. There are a lot of properties that aren't systematically captured in the periodic table that are chemically important, but we see a lot of interesting trends. The versions that are most often around today capture trends that we think are really important to the central activities of chemistry, which are chemical reactions. I wonder if part of the issue is we don't have the central activities of nano as well-defined yet. We have some application-driven clarification but – I think I asked you this question maybe five or six years ago with within a particular view in mind. I'll just ask the question again: is there going to be a periodic table of nanoparticles? Why or why not?

MILLSTONE: Now, I would say no. I don't think we can conceive of enough dimensions to make one. You know in all of the different types of schemes that we've talked about, I'm not even sure we know what's important. I haven't really studied the history of Mendeleev, I think I probably should, because of course there are tons of things he didn't know and it still worked out great. I could be missing the forest for the trees. I should really go back and look at what stood out to him and see if there are analogies. I mean of course that's not the way to make scientific breakthroughs, but still I'd be curious.

BURSTEN: I think using history as a tool for doing science is probably an undervalued resource.

MILLSTONE: It really probably is. Not probably. It definitely is. When you're teaching general chemistry like I am now, you can look at some of the experiments that people did to uncover massively fundamental aspects of how we understand matter, it's unbelievable. Like, I passed a ray of particles through a magnet, saw that it divided, and discovered that electrons have spin? Whoa. The schematic is pretty straightforward, the students could set that up, but the physical knowledge to interpret that experiment and the creative power behind thinking that, oh this means that these are two different particles and they have properties that are as if these particles have spin – incredible.

BURSTEN: OK, the big thing I want to make sure we talk about before we finish is the four-part characterization of nanomaterials that you use, this characterization scheme of composition, size, shape, and surface chemistry.[2] Walk us through how this scheme came about. Do nanoscientists talk about classifying nanomaterials in this way in textbooks? How do your colleagues talk about it? Why have you landed on that particular four-part scheme, and what do you think it gets you?

MILLSTONE: The four-part idea is empirically borne out of what we have observed to influence the properties. We know from chemistry in general that the composition matters, so we know if we have cadmium selenide versus copper selenide, the properties will change. Composition is where we start as chemists. If you can change the composition, you're going to change the chemical and physical properties. Then what we noticed first in nano was that if you change the size of something that has the same composition now, unlike we thought before, now the size has an influence on the observed optical and electronic properties, whatever those might be.

BURSTEN: And that's not just macroscopic or bulk materials to nanomaterials, right, that's even within nano?

MILLSTONE: Right, so if you have a 50 nm particle, it's different from a 20 nm particle, and then there will be some size, on the high end where you'll stop seeing size-dependent properties. Maybe you'll have a 300 nm particle and a 400 nm particle and they're going to be basically the same. Where that line is will depend on the material, but you'll have some size at which now you don't see size-dependent properties anymore.

As we grew synthetically, as we had new capabilities synthetically, we started to see that changing the shape, even for something that has the same total volume, produces different properties. Then, over the last 10 years, we realized that surfaces even on these materials can also change properties, so you can have two different particles both 5 nm in diameter, and if you change the composition of the ligands binding to the surface, then you'll change the luminescent properties of the material, for example. That's true in metals, and it is true in semiconductors too. So we realized about size and then shape and then that surface chemistry can also have the same impact.

There's actually a really nice example I like to use from Brandi Cossairt's group at the University of Washington. She did a really nice study on indium phosphide particles, changing the atom content at the surface and showing really big changes in luminescence. It looks like a stoplight, just from surface chemistry changes.

This is a classic you-and-me conversation. When I think about why these four features, it's also the case that those are the synthetic parameters we can control, so those are the synthetic parameters that we've seen influence chemical and physical properties. To go any deeper than those categories, we need different synthetic capabilities. So, for example, something that is not captured in there but we know makes a difference is something that I call composition architecture. This is, for example, the difference between a core-shell structure, where one metal is encased by another, versus a Janus particle, where the two metals sit side by side. The two architectures can have the same total metal atom content of each metal but be distributed differently in the particle. That creates different properties. That isn't captured in the scheme but it is really important, and people have been trying to control it. The time period over which people have been trying to control composition architecture overlaps enormously with the use of these categories, so we've known that the scheme doesn't completely capture what we're talking about for a while. So there's at least one thing that it doesn't capture that we can synthetically control and it has a big impact. There are probably other things.

BURSTEN: Synthetic considerations, like whether it's seed-mediated growth or other types of growth, also seem escape the scheme, or do you roll that into the final architecture?

MILLSTONE: That's right.

BURSTEN: I could envision a classification scheme where that's part of it.

MILLSTONE: I could envision something like that too. This classification scheme is final architecture, path-independent. For shape, for example, is shape the right metric to put there? We put shape but, you know, should it be aspect ratio? What is actually the figure of merit we should be using in terms of shape? Size has a figure of merit. Surface chemistry is nebulous but for the right reasons. Shape, probably, we could push on a little bit and get more quantitative, but we also have a pretty good geometric understand of shape so it's fine.

BURSTEN: Composition is nebulous too, for alloy particles.

MILLSTONE: Absolutely. Composition architecture is where I would encapsulate that problem. We used not to think of composition as a topic of study in nanomaterials, because we would take for granted that you change the composition you change the properties. But composition now becomes, when we are talking about properties that change on the nanoscale, now we should probably replace composition with composition architecture. Composition is obviously going to change with the properties – that's true at any length scale. Composition architecture is a synthetically targetable feature of the nanomaterial and will impact properties.

The scheme is mainly derived from the fact that these are the things we can empirically control, and therefore we've observed them to have a significant influence on chemical and physical properties. And it does withstand some non-trivial scrutiny. Let's say I could control the particle atom by atom: I would still start to see just gradations within each one of those families. On surface chemistry, is ligand density or ligand arrangement more important? Is surface atom population or is the organic molecule on the surface more important? On shape, is the aspect ratio more important or the actual geometric shape more important? The thing about the gradations, like is it aspect ratio or is it shape, it's hard to start homing in any more specifically because shape influences plasmonic features and so does size. Which one has more impact and what's the impact? Is there an interdependence? That's where it starts to get hard, really, really hard, very quickly. Am I getting at the question you're asking?

BURSTEN: I think so. You're definitely getting toward it. Let's do a little background check-in. This is a four-part scheme. Is this what shows up in most nano textbooks as the figure of merit for understanding nanoparticle characterization?

MILLSTONE: I'll be totally honest with you. I made up the scheme in 2007 for my dissertation defense. But I do see the spirit of the scheme in a lot of places now. I'm not saying it came from me. I do see it a lot of places now. I would say people in general quote this idea.

BURSTEN: There are nanoparticle ontologies in databases now, where you categorize nanoparticles by tagging them.[3] Some of the tags are synthetic-process dependent. I want to hear a little bit more about how other people thought about this. Do you think people who are doing physical stuff for synthesis, as opposed to colloidal synthesis, or seeing different characteristics, because they got different kinds of synthetic control?

MILLSTONE: I'll give you an example that was not encapsulated there that will influence the physical and chemical properties. That's the dispersity of the sample. The classification scheme we're talking about depends on having total uniformity. If we're making a nanoparticle in the way of solid particles suspended in liquid media, then the way we're making those materials, you cannot achieve an individual peak. Every particle is not the same. So the chemical and physical properties that you measure obviously are influenced by the fact that for the most part you're not measuring single particle properties. Dispersity would be an additional influence on each feature of the classification scheme, each one of the chemical and physical properties that you would observe. But if we want to get down to the essentialist part we wouldn't consider the dispersity.

This ties in to a comment I wanted to make about path-dependent properties in a classification scheme, whether we want to take the synthetic path into account. I have to tell you I would say, "No," unless dispersity is something we're going to include. If the path matters, based on how I think about material properties, if they are dependent on structure, then I would say whatever structure you end up with, that's what determines chemical and physical properties. If we saw that we could get an identical structure but it had different properties based on the path, then most people would look for something residual in the structure reflecting that path. I probably wouldn't include the synthetic route unless dispersity was an issue because the route will influence the dispersity even if your average particle morphology is the same.

As I said, the dispersity could be in any one of the parameters, could be in the surface chemistry, could be in the shape or the size, and in fact it is in all three. For example, if I have two different sizes and then for each size, two different shapes, or I have two different shapes, and each shape has two different sizes, and each one of those has different surface chemistry, it's just compounding.

BURSTEN: Can we talk a bit about the relationship between the four? You were gesturing at this a little earlier that composition tends to restrict the other three.

MILLSTONE: That's right, yeah.

BURSTEN: But it's not going to be a sort of easy flowchart taxonomy kind of thing. We started talking about this a little bit earlier, the idea that whether or not we could aim for something like a periodic table for nano. There's no periodic table for materials, and materials are the other comparison case that we look at when we're thinking about how to classify in nano. Do you remember the Ashby diagrams, these materials charts with comparing properties like hardness and density of these pockets of like ceramics and plastics and things?

MILLSTONE: Yeah.

BURSTEN: That seems to me a bit closer to where we're going to go with nano. It's not going to be a simple two-feature thing.

MILLSTONE: Yes.

BURSTEN: I wonder if you can frame some of the relationships between the four.

MILLSTONE: There are two obvious ones that I can point out that won't be controversial. First, composition and shape. If you change the composition, you change the low-energy crystallographic structures. Some metals have based-centered lattices and some have face-centered lattices, for instance. Changing the low-energy lattices by definition then changes what crystal facets are exposed, which changes what are the low-energy shapes you might get. That also changes the low-energy arrangements of ligands that you might get for any given chemical identity of the ligand.

Likewise, if you look at what ligands are going to be on the surface, the chemical identity of the surface, regardless of the crystal facet, has a big influence on that. For a given composition of crystal and particle, there's going to be some series of shapes that are more favorable than others. That's going to mean that although they may not be impossible, it will be much, much more likely for you to observe this particular suite of shapes than another particular suite. Likewise, different ligand architectures and types of ligands can be stably attached. Composition has an influence on all of those different parameters.

Shape also has a relationship with surface chemistry. Even for the same composition, different shapes are going to present different crystal facets, which will influence ligand arrangement, and we don't really have to get any more creative than that to point out a very impactful consequence of shape. That's a very obvious one.

Size and shape, there are some shapes that will grow out over time so you can get high-energy intermediates, but they're not stable, and so on. If you have enough precursor you're going to grow out the shapes, stuff like that, but that's not the kind of relationship we're talking about. These are not kinetic relationships; these are thermodynamic-type relationships. The inter-relations between them are non-trivial, although for some reason those are not standing out to me as the big barriers in classification.

BURSTEN: All four of these properties are some variant of a feature of nanoparticle structure. Composition architecture is also a very structural property: it's trying to get at a fine-grained internal structure that you don't capture through any of the other four. Here's a philosopher's question: why don't we just classify nanoparticles by structure?

MILLSTONE: Great question. You mean, why don't I just say what is the size? What is the shape? What is the surface chemistry?

BURSTEN: Not quite. I mean not even that. Why can't we roll the classification scheme all into one description of, "The structure of nanoparticle X is Y," and do some complicated naming scheme like organic chemistry, that's going to tell me all of the molecular-geometric relationships among each of the atoms, or do something where we just basically draw diagrams every time, which is what the complex organic chemistry naming schemes are anyways, linguistic encoding of diagrams. Why does it make sense to stop along the way at these four different structural features as opposed to just saying it's all structure?

MILLSTONE: Because the second I start to try to dig deeper and put a name to it, it becomes insanely complex. Surface chemistry is the best example of this. I'll try to say, "What structural features of surface chemistry do I need to include in the identifier for this nanoparticle?" Let's assume that analytically I could measure all of these. I may or may not need to include the atomic ratio and atomic arrangement of all the surface atoms. I may or may not need to include the number and relative arrangement of all of the surface ligands, whatever those might be. I need to also characterize their binding motif, because a thiol can bind multiple ways to a surface. All of those things I need to specify. In some particles, all these parameters are going to matter, and in some particles, they're not going to matter.

Maybe if we had the analytical tools to answer each one of those questions, the tractability of defining the surface chemistry would get easier. In fact, it's obvious that it would. If I could look at these things routinely, I could figure out what elements of the surface chemistry I need to define. Also, all those things I just said are per crystal facet.

Right now, we cannot measure any of those things routinely. Even when we can access or get close to those parameters, they are an average over the entire sample. We don't know particle-specific information, and so, I like where you're going, which is that theoretically it is possible to potentially define these by structure alone. At least that's what I would be trying to do, but I simply can't measure.

BURSTEN: I actually think you were trying to do that when we first met seven years ago, but I don't think you are trying to do that anymore. I want to get at that, so I'm going to give you an analogy. In the periodic table, people were happy with the Mendeleev table by the time we understood nuclear structure. When we got the nuclear structure we realized that there are isotopes. We could have rewritten the periodic table and said each isotope is its own element and done the individuation at that level, because there are physically and chemically different properties when we get two different isotopes, and we know that those physically and chemically different properties matter. They make a difference to what can react, they make a difference to biology, they make a difference to chemistry, they make a difference to physical properties like melting and boiling points. But we didn't revise our periodic table to be the Periodic Table of Isotopes. It's still the Periodic Table of "things that have the same number of protons." Similarly, we could have done ions. We could have said every ion is its own species because ions are going to act very different, chemically.

So, in a similar way, I could see either a finer-grained or a coarser-grained version of the structure story that you're telling. Either would get us away from this sort of four-part scheme. You went exactly to one of the examples I was thinking of, which is just like this super, super fine-grained analysis of surface chemistry for each particle for each crystal facet. Each concentration of precursor would give you a slightly different ligand dispersity on the surface. Why is this four-part scheme the level that's making sense to you?

MILLSTONE: Let me go back to the analogy you were making with the periodic table. With the ions, it's true, we could have done it differently. We could have invented a new classification scheme. But it turns out that periodic trends help you a lot to understand ions too. And the same with isotopes. It turns out that the connecting feature of isotopes is these things like isotope effects, where the heavier atom slows down kinetics or whatever. With isotopes, it seems like it's not the periodicity so much as the fact that it has extra neutrons. You're right that the classification would have looked really different from the periodic table, but the periodic table still gives us a lot. It even comments on those other species that aren't explicitly treated.

For nanoparticles, I think you're pointing in a really interesting direction. What is the "ion," what is the "isotope" of nanoparticles? I think we need a lot more analytical work before we know. But I think you're right, there are going to be more detailed levels of each one of these that is going to arise to be that may enable a classification scheme that is like organic chemistry. Right now it's really analytical tool-dependent, because we can't figure out these aspects if we don't know if they're important. We don't know, for instance, if you say, "I have a thiol-protected gold cluster with an average size of 144 atoms," if that tells you everything you need to know. It might. But we can't create the alternative to test.

BURSTEN: I think the thing you get out of an organic type scheme or even out of the periodic table, that you don't get out of the kind of material clusters classification system, is a well-defined set of unknowns. If you're looking at the periodic table, you know there's a hole at 117, you know what 117 should look like if you could ever find it. It suggests experiments, it suggests a hole in the model. If I know that I have a carbon backbone and I know that I have three functional groups and four carbon sites, then I know the number of permutations I can run through to stick on the functional groups. I know that I can compare the chemistry of the functional groups to see which ones are going to be thermodynamically impossible, which ones are going to be harder or easier or lower energy. Materials classification schemes, where you're clustering things by properties, that is extremely useful and reveals these pockets of similar materials, but it doesn't suggest known unknowns in the same way. I wonder if that's just a feature of the part of the timeline we're on for nanoscience, or if nanoscience is genuinely like materials. It's not like we know that there are 18 possible kinds of steel out there and we've made 16 of them. Steel changes in a much more fluid way and we can use certain assembly techniques, we can manipulate the alloy structure in the initial mixing, we can manipulate the grain structure and orientation on these almost continuous levels. Nano is right at this interface, and I'm wondering if we should be aiming for something like an organic structure if that's going to limit us.

MILLSTONE: You're 100% right. You know, you've got a couple of things from this that you're going to leave home with. That's the conclusion that I'm going to leave home with. In the beginning, when nanochemists were defining nanoparticle reactions, people were embarrassed to talk about how stirring rate influenced what they got, and now we know it's important and we report it and you're considered a fraud if you don't.

BURSTEN: Can you explain that example? What happened when people weren't reporting stirring rates and are now reporting stirring rates?

MILLSTONE: What was happening in the early days of nanosynthesis was that people would notice that mundane aspects of the synthesis would have a serious influence on the particle outcome, like stirring rate. How fast your stir would influence the homogeneity of your nanoparticle batch. Or you would see that it would influence the shapes you get. In a case where you'd stir differently (maybe still quickly, but not *as* quickly or using a different stir-bar), you'd have a hiccup in the stirring and you would see a lot of different shapes. Same thing with injection rate. If you would inject one of the reagents slowly, you would see a totally different thing than if you injected it quickly. Now, with the stirring rate, if you're an organic chemist, whether you stir fast or a little bit slower, as long as you're stirring, typically, it's going to be fine. Maybe there are some reactions that are super sensitive, but in general it's not really a big thing.

BURSTEN: The electrons are going to get there.

MILLSTONE: Yeah, it usually will work fine. Unless you're talking about a polymerization, it really doesn't matter all that much whether your stir-plate is set to rate 8 or 9. But in nanomaterials, in the first reports of the different syntheses, people were having a huge problem reproducing anybody's synthesis methods. There was a lot of heterogeneity. Sometimes it would work, sometimes it wouldn't work. Over time, there were a lot of different factors in the synthesis that started to reveal themselves as important, which we wouldn't have thought of previously. Things like impurities in the reagents had huge impacts on the final morphology, things like stirring rates, how long it took you to inject a reagent. It became clear that those things did have an impact on the morphology. The reason they did wasn't voodoo. It turns out that that's how nucleation works, and there was theory to describe and predict that if you weren't stirring properly you're going to have heterogeneous growth, because you're going to have a little bit of nucleation, those guys will grow, and then something else over here will nucleate 10 seconds later. As soon as we recognized that this is important, it became obvious that it fit with what we know about colloid growth and nucleation. People started reporting nano-materials syntheses differently.

With thinking about nanoparticle classification like materials classification, and the synthetic implications, it's going to be the exact same thing. People have this preconceived notion of what a mechanism for nanoparticles is going to look like, and it doesn't look anything like that. That's awesome.

BURSTEN: Yes! OK, I love this too. Unsurprisingly, this is where I want to go with it. I want to say it's not an in-principle thing, it's not that we just don't have the analytical tools yet, it's not just that if we could do this atom-by-atom that would be better, but we have these sort of computational stopgaps. It's that nano is like materials. We haven't defined our unknowns and we're not going to. That's not the game here.

MILLSTONE: That's right. This is exactly it.

Notes

1 For the history of this naming convention, see Chapter 2.
2 For an extended philosophical discussion of this scheme, see (Bursten, 2018) and Chapter 7.
3 For a discussion of ontologies, see Chapter 9.

References

Bursten, J. (2015) *Surfaces, Scales, and Synthesis: Scientific Reasoning at the Nanoscale.* (Doctoral dissertation) University of Pittsburgh, Pittsburgh, PA.
———. (2016) Nano on Reflection. *Nature Nanotechnology.* 11(10), 828.
———. (2018) Smaller Than a Breadbox: Scale and Natural Kinds. *British Journal for the Philosophy of Science.* 69(1), 1–23.
Bursten, J., Hartmann, M. & Millstone, J. (2016) Conceptual Analysis for Nanoscience. *The Journal of Physical Chemistry Letters.* 7(10), 1917–1918.

9 Categorization of nanomaterials
Tools for being precise

John Rumble, Jr.

Introduction

As discussed in other chapters in this volume, the problem of aggregating items
into natural and unnatural kinds – that is, grouping items according to defined
criteria – is old and has been a topic of discussion by philosophers through the
ages (for a historical review, see Bowker and Star, 2000). With the advent of the
scientific revolution 500 years ago and the more recent Information Revolution
(Isaacson, 2014), such aggregation, or as we shall refer to it – categorization –
has become increasingly important in science, from both a theoretical and a
practical point of view. From a practical point of view, categorization in science
is an enabler that allows us to recognize similarity in objects such that understanding the behavior and properties of one object allows understanding of the
behavior and properties of a similar object. From a theoretical point of view,
categorization in science allows us to provide explanations for behaviors and
properties, either as a correlation – behaviors or properties in similar objects
can be correlated with one or more features of those objects, or as a cause and
effect – behaviors or properties of similar objects are cause by specific features
of those objects.

Before proceeding, we need to define what we mean by the term categorization
and how it applies to physical objects, which, in this chapter, refers to physical
and virtual nanomaterials.

By *categorization*, we mean the grouping of items according to defined criteria. Ideally, the criteria are well-defined and measurable. Much has been written
about whether a categorization system should result in non-overlapping categories, apply to every object in a set (no item that cannot not be categorized), or have
no ambiguity (e.g., sharp distinctions between categories, such as small, medium,
and large).

By *physical objects*, we mean a tangible object composed of atoms and molecules, or the particles that make up atoms and molecules. We include virtual
descriptions of tangible objects as long as they are realizable. Excluded are theories (e.g., theory of relativity), abstract concepts (e.g., beauty, health), mathematics (e.g., numbers, equations), and intangible ideas (e.g., time-warp). Physical
objects are of three types: materials, which are collections of one or more atoms

or molecules; devices, which are one or more materials designed to perform a function (e.g., movement, transport of electrical current), and systems, which are multiple devices designed to provide higher level functionality (e.g., transportation, communication).

Humans have always used categorization, if only as a survival skill to avoid predators and identify food. Categorization was important to early scientists as a mechanism to organizing knowledge. The scientific revolution of the 16th and 17th centuries brought a new emphasis on using measurements to quantify our knowledge of nature and thereby improve our ability to categorize. Measurement techniques were not only systematized and codified, but also allowed detection of property differences on finer and finer scales. This new measurement capability resulted in a new awareness of the importance of categorization for coping with the large volume of data being generated by modern science.

Over time, it became clear that the possibility of making every conceivable measurement was remote. As science moved into the 19th and 20th centuries, the number of items of scientific interest – from animal and plant species to atoms to molecules, to materials to astronomical objects to micro-organisms – grew ever larger. Then over time, the problem evolved from making an impossible number of measurement possibilities to predicting measurement results based on fundamental and empirical theories. The use of categorization as a necessary tool for predictive science became a major driving force in reducing this measurement burden (Martens and Martens, 2001).

The specific physical objects that are considered in this chapter are nanomaterials, that is, materials that have one or more external dimensions on the nanoscale. The nanoscale is defined as approximately one to approximately 100 nanometers. A more detailed discussion of definitions of nanomaterials is given later.

The structure of this article is as follows. In the next section, we briefly discuss the importance of categorization in science, and especially for nanomaterials. Following that, we explore the general aspects of the categorization of nanomaterials. Next, we detail a nanomaterials description system developed by the CODATA Working Group on Nanomaterials and then discuss additional issues concerning the categorization of nanomaterials, before presenting some conclusions.

Importance of categorization of nanomaterials

Over the last twenty years, nanotechnology research has led to new capabilities to conceptualize, make, measure, and exploit nanomaterials (Vance et al., 2015; Xia, 2014; Roco, 2011). Nanomaterials exhibit new or improved properties because of (1) their atomic and molecular make-up, (2) unique quantum effects based on size, and (3) surface characteristics. The variety and nature of nanomaterials is large, and their applications are equally numerous. Their commercial uses are almost limitless, and variations among them subtle and often elusive.

As for any other set of scientific objects, categorization of nanomaterials is important. We can identify three different types of goals for categorization of nanomaterials. One set is scientific, a second set is application oriented, and a third set is methodological.

Scientific goals

The scientific goals for categorizing nanomaterials include (1) improving the predictability of nanomaterial properties and functionality, (2) providing ideas and clues for designing new nanomaterials with desired properties, and (3) understanding the life cycles and mechanisms of actions of nanomaterials.

First, a prime motivation focuses on the predictability of properties and functionality for similar nanomaterials. In the case of small molecules, organic chemists in the first half of the 20th century used the identification of functional groups, for example, alcohols, aldehydes, and ketones, to categorize reactions and synthesis pathways. The latter decades of the 20th century saw the development of numerous quantitative structure-activity relationships (QSAR), especially focused on pharmaceutical activity (Hansch and Fujita, 1964; Fujita et al., 1964; Gramatica, 2007; Mannhold et al., 2008). QSAR involves dividing comprehensive small-molecule libraries into smaller subsets of similar objects based on structural or compositional features. Then representative members of each subset could be tested for different types of pharmaceutical activity. When activity was found, other members of the category could be more intensively tested to identify related but more active objects.

The same reasoning applies to nanomaterials. Once a specific nanomaterial feature, for example, chemical composition, shape, and aspect ratio has been found to correlate with some desired property or functionality, other nanomaterials with similar features can be explored. High-throughput screening techniques now provide powerful tools to explore large libraries of nanomaterials easily (Nel et al., 2012; Thomas et al., 2011).

Next, consider the need for clues to new materials designs. Predictability relies on similarity, but categorization also provides a capability to identify outliers and anomalous behavior that generate ideas for new avenues of exploration. Such clues often are critical in providing insights into new ways to look at the behavior of objects, in this case, nanomaterials. Why doesn't this nanomaterial behave as expected? What causes it to have this different property? The fact that something does not have an expected result is common in science and is often the basis for serendipitous discoveries.

Finally, consider the need to understand the life cycles and mechanisms of actions of nanomaterials. Correlation of properties and functionalities with features is not enough; scientists want to understand the cause of effect. For nanomaterials, the fact that they are so reactive adds complications beyond those encountered with small molecules and bulk materials. In addition to identifying which feature of a nanomaterial causes a specific effect, there are the complications associated with changes to composition, structure, and other features during the lifetime of a nanomaterial. (On context-sensitive classification over the lifetime of an object, see also Chapter 4 and Chapter 7.) This changing nature of a nanomaterial through its life cycle will be discussed ahead. Here it is important to note that as researchers work to elucidate a mechanism of action – how does a feature lead to functionality – the fact the nanomaterial is likely to change through the testing process complicates the process.

The role of categorization here is to posit that similar nanomaterials will have similar life cycles, given the same exposures and post-manufacture treatment, and therefore have similar mechanisms of actions.

Application goals

Nanomaterials applications (e.g., electronic components or drug delivery) define two additional goals for categorization of nanomaterials: reducing the cost of testing nanomaterials with similar features, and predicting the impact of nanomaterials on living systems.

Toward the former, consider that while the number of nanomaterials does not yet compare to the number of inorganic (about 160,000) (FIZ Karlsruhe, 2016) or organic molecules (approaching one million) ("Cambridge Crystallographic Data Centre", 2016), it can be instructive to study principles of classification in those domains in order to understand the challenges of classifying nanomaterials. Looking at the case of small molecules, the fact remains that very few of them have more than their chemical formula and structure measured. Data for fundamental physical properties such as melting points, boiling points, heat of formation, and so forth exist for only a fraction of known compounds. Even though tests for determination of these physical properties are quite standardized, they still require an investment of time and money that is often lacking without some motivating factor such as commercial importance. In many cases, these properties can be estimated from empirical equations parameterized by measurement results from "similar" molecules, where similarity may be based on structure or component moieties. It also is possible to classify small molecules without necessarily measuring *all* the physical properties.

A parallel situation exists for nanomaterials. Researchers often prepare many related versions of a basic nanomaterial, but lack motivation for measuring any but the basic structural or compositional properties. One driving force for categorization of nanomaterials is the hope that estimation techniques can be developed comparative to those used for small molecules (Hansen et al., 2014).

Nanomaterials have many small variations that greatly increase their number and variety. This complexity provides both opportunities and challenges. The variations arise for three major reasons. First, producing exactly uniform nanomaterials is very difficult, with most production processes resulting in a range of sizes and compositions, especially for mass-produced nanomaterials. Second, the reactivity of nanomaterial surfaces means that even in benign environments, aggregation, agglomeration, and reactions with the inevitable impurities often result in size and compositional changes. Third, in complex media, such as biological fluids and environmental systems, nanomaterials are easily coated, often going through a series of coatings as they are exposed to different substances. Often commercial nanomaterials are deliberately coated to achieve a specific property, for example, wettability or dispersability, and then coated again by material in the media of application.

One result of the ease of coating is the ambiguity of which nanomaterial variation is the cause of a particular behavior or property. Understanding and controlling the life cycle of a nanomaterial is the subject of much research. While we often envision nanomaterials as perfect spheres, in reality, they come in many shapes and with many imperfections. Over time, individual atoms, molecules, or small chunks can break away, leaving irregular surfaces, edges, or asymmetrical shapes that are difficult to characterize.

An overarching concern is the ambiguity of when an instance of a nanomaterial is simply a small variation and when it is a "new" nanomaterial. For uses which embed a nanomaterial within a bulk, or larger, material structure, this is less of a concern. For uses in which a nanomaterial is able to move freely, these concerns are more challenging.

Relatedly, there are challenges in predicting the impact of nanomaterials on living systems. Understanding what impact nanomaterials have on living systems is of considerable interest today. Many concerns have been raised about the potential negative effects nanomaterials might have on humans, other animals, plants, and other living things. The testing for possible toxicity is a time-consuming and laborious process. With new nanomaterials constantly being developed, and given that manufacturing "well-known" nanomaterials produces a range of size, shapes, and other features, testing for toxicity and other properties on all variations is close to impossible.

Categorization offers two possible solutions to this problem. First, some categories of nanomaterials can be shown to have no toxicity effects; second, other categories will show possible effects. For the former, new members of a class can then be quickly screened to show that their behavior presents no problems. This screening could be accomplished by identifying one or a few key tests that validate a "new" or "modified" nanomaterial indeed is a member of a preexisting class. This approach is now proposed by the EPA using properties such as zeta potential, specific surface area, dispersion stability, or surface reactivity as potential mechanisms to identify a new nanoform (EPA, 2018). In the latter, new members will have to be tested more thoroughly. In both cases, categorization supports predictability in addressing a serious barrier to commercialization of nanomaterials (Hansen et al., 2014).

Methodological goals

Before we consider specific categorization approaches, it is useful to discuss three goals that provide a more methodological framework for whatever approach is used: uniqueness and equivalency, causality, and context.

First, the foundation of any categorization approach is a system that describes objects uniquely and supports equivalency (Rumble and Freiman, 2012). In terms of nanomaterials, this means that there is a system that allows clear description of a specific nanomaterial so that it is clearly distinguishable from all other nanomaterials. This establishes the uniqueness of a nanomaterial. Similarly, once uniqueness is established, the equivalency of two nanomaterials can be determined, which is the case when every feature of one nanomaterial occurs in another nanomaterial. Uniqueness allows us to know exactly which nanomaterial we are describing. Equivalency allows us to combine measurements of two equivalent nanomaterials into larger data sets.

Second, consider causality. A causality model must link the categorization feature or property with the outcome (impact). Heuristic and correlative models are not causality; one must be able to control the feature and predict the outcome to demonstrate causality. One of the most important discussions of the difference between association (correlation) and causality is that given by Hill (Hill, 1965; see also Phillips and Goodman, 2004; Woodward, 2003). Hill presented nine criteria for differentiating between causality and correlation. While these criteria were developed in the context of epidemiological studies of disease, they are applicable to other situations, including categorization of nanomaterials. Six of the criteria are summarized in Table 9.1.

With the causality criteria established, the last methodological goal is that of context, specifically the need for a clearly defined context of investigation or application. The scientific goals outlined in the first part of this section are part of larger contexts for categorization of nanomaterials, which include contexts such as commercialization, regulation, pharmaceutical activity, and many more. Different categorization systems for nanomaterials are possible depending on the context, and a system suitable for one context may or may not be useful in another context. In fact, the features involved in causality for one context may not matter in another.

The context-dependence of nanomaterials categorization is important in two ways. First, data gathered for establishing categorization in one context may not be sufficient or useful for establishing categorization in another context. Therefore, the ability to integrate data from different contexts is important to have the largest possible data sets available. Second, categorization of nanomaterials with respect to their potential and actual biological effects should not obscure the need for other contexts. Consequently, characterization of nanomaterials for toxicity testing, for instance, should include measuring as wide as possible array of properties that could be useful outside of that context.

Table 9.1 Criteria for determining causality

Criteria	Comments
Strength	A small association does not mean that there is not a causal effect, though the larger the association, the more likely that it is causal
Consistency	Consistent findings observed by different persons in different places with different samples strengthens the likelihood of an effect
Specificity	The more specific an association between a factor and an effect is, the bigger the probability of a causal relationship
Temporality	The effect has to occur after the cause (and if there is an expected delay between the cause and expected effect, then the effect must occur after that delay)
Biological gradient	Greater exposure should generally lead to greater incidence of the effect; however, in some cases, the mere presence of the factor can trigger the effect
Plausibility	A plausible mechanism between cause and effect is helpful (but Hill noted that knowledge of the mechanism may be limited by current knowledge)

Approaches for categorization of nanomaterials

In this section, we briefly describe several possible approaches used to categorize nanomaterials. Space does not permit an extensive review of any specific approach, but references provide further details.

Four views of nanomaterials categorization

There are many ways to approach categorization of nanomaterials, but here we adapt a four-view scheme, proposed by Sellers et al. (2008) as schematically shown in Figure 9.1.

Each categorization view, or approach, can include many different types of information depending on the context. Later on in this chapter, we discuss in more detail issues related to the definition of a nanomaterial as well as certain aspects of the information included in each component. In Table 9.2, we identify some of the information used in each categorization approach. The lists are not exhaustive.

Implementing nanomaterial categorization

For small molecules, comprehensive, complete, and systematic data relevant to categorization, that is, complete specification of composition, structure, bonding, and stereochemistry, are available in public databases, such as those for

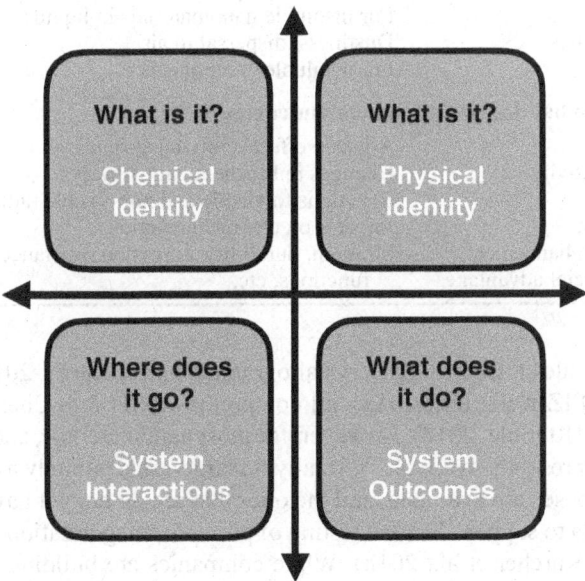

Figure 9.1 The four-view scheme for nanomaterials categorization

Table 9.2 Information used to categorize nanomaterials from different viewpoints

Information type	Comments
Viewpoint 1: What is it? Chemical Identity	
Overall chemical composition	The general chemical composition of the nanomaterial
Chemical composition of different internal structures	If the nanomaterial has more than one internal structure (shells, layers, etc.), the chemical composition of each
Crystallographic structure of each structure	Different phases have different crystallographic structures
Impurities	If the nanomaterial is an aggregate of many nano-objects, which chemical impurities are present
Surface composition	Chemical composition of surface
Viewpoint 2: What is it? Physical Identity	
Size	Physical dimensions of nanomaterials (mean, average, actual, etc.)
Size distribution	The range and occurrence of different size nano-objects within the nanomaterial
Shape	The 3-D or 2-D shape, or projection thereof
Physical structure	The assemblage of structure components
Structural features	Holes, protuberances, etc.
Structural characteristics	Regularity of structural features, aspect ratios (length to width), etc.
Charge distribution	Location of positive and negative charge on surfaces
Viewpoint 3: Where does it go? System Interaction	
Solubility	In water and other liquids
Hydrophobicity	Attraction or repulsion of water molecules
Dispensability	For insoluble nanomaterials in liquids
Aerosol properties	Dustiness, dispersal in air
Dissolution	For insoluble components
Viewpoint 4: What does it do? System Outcomes	
Toxicity	Adverse effect on living systems
Biological reactivity	Changes to biochemical pathways
Photoreactivity	Reactions to visible and non-visible radiation
Catalytic action	Impacts on chemical reactions
Functionality enhancement and commercial advantages	Strength, durability, corrosion resistance, electronic functions, etc.

organic molecules ("Cambridge Crystallographic Data Centre", 2016), inorganic compounds (FIZ Karlsruhe, 2016), all compounds (NIH Pub Chem, 2018), and property data (Rumble, 2018). However, for most nanomaterials, analogous information is not presently available. Virtually no large-scale publicly available nanomaterial databases are available, and the ones that are do not yet have a sufficient amount of data to support detailed testing of possible categorization schemes (see, for example, Karcher et al., 2018). While companies are building internal databases of the nanomaterials they produce, they are not yet broad enough to cover the wide variety of nanomaterials in existence.

Perhaps more important, the lack of precision in describing nanomaterials well enough to correlate specific features with important properties has hindered establishing cause-and-effect relationships between observed nanomaterial features and their properties.

Describing nanomaterials accurately

One of the critical factors required to meet the goals for nanomaterials categorization is a robust method of describing nanomaterials accurately. If nanomaterials are to be put into categories, regardless of what scheme is used, the categorizing features must be well-defined and described accurately. While standards development organizations such as International Standards Organization (ISO) Technical Committee 229 Nanotechnologies (ISO TC 229, 2016) and ASTM International Committee E56 on Nanotechnology ("ASTM E56 on Nanotechnology", 2017), as well as the Organisation for Economic Co-operation and Development (OECD) Working Party on Manufactured Nanomaterials (Organisation for Economic Co-Operation and Development, 2016) have put in place a variety of partial description recommendations, those schemes are not adequate. More specifically, they have not created systematic and cohesive information models that detail all the features of nanomaterials, reflecting the state of new knowledge continually emerging in nanomaterial research.

The CODATA Uniform Description System of materials on the nanoscale

In response to this need, a joint working group ("CODATA Nanomaterials WG", 2017) of CODATA, the International Council for Science (ICSU): Committee on Data for Science and Technology ("CODATA", 2017) and VAMAS, an international pre-standardization organization concerned with materials test methods ("VAMAS", 2017) developed a systematic approach to describing nanomaterials accurately. This system is called the Uniform Description System for Materials on the Nanoscale (UDS) (Rumble et al., 2014).

The working group was set up to foster the development of a uniform description system for nanomaterials to address the growing diversity and complexity of nanomaterials being developed and commercialized. The group included representatives from virtually every scientific and technical discipline involved in the development and use of nanomaterials, including physics, chemistry, materials science, pharmacology, toxicology, medicine, ecology, environmental science, nutrition, food science, crystallography, engineering, and more. Many international scientific unions have actively participated. An initial draft Framework for the UDS was developed and made available for further comment and modification. Version 2.0 has been available since 2016 and may be downloaded at https://zenodo.org/record/56720.

The approach taken by the UDS has been to identify the broad types of information that are used throughout the nanomaterials community to describe a nanomaterial as completely as possible. As discussed earlier, the two goals are (1) to establish the uniqueness of a nanomaterial so it is clear which nanomaterial is being described and (2) to allow the establishment of the equivalency of two nanomaterials to whatever level is desired. The UDS uses natural language so that most of the terms and concepts used in the description system are readily understandable to the scientists, technologists, and lay persons involved in nanotechnology. The description system can be used by different user communities, including informatics experts who design and implement data and information resources using the latest informatics tools such as information modeling, ontologies, and semantic web technology.

The basic premise behind the UDS is that a nanomaterial cannot be uniquely specified by a simple or systematic name in the manner of chemical nomenclature. (On chemical nomenclature, see also Chapter 2; on the limits of nomenclature in nanomaterials, see also Chapter 8.) Further, the description systems developed for metals, alloys, ceramics, polymers, and composites are also in an inadequate state for nanomaterials because of size, surface, shape, and other effects that significantly influence their properties. Simplistic terms such as "carbon nanotubes" or "quantum dot" convey important information, but accurate and complete identification of a specific nanomaterial requires more details. Instead, for complete specificity, all relevant information categories need to be used. Many situations require this level of specificity including the development of regulations, standards, purchasing, and testing.

Types of nanomaterials

Throughout this chapter, the term *nanomaterials* is used to mean *materials on the nanoscale*. While a variety of definitions of nanomaterials exist, two major international standard definitions have been adopted. The definition in this chapter is intended to be compatible with both standard definitions.

The ISO TC229 definition of a ***nanomaterial*** is as follows (ISO TS, 2007, p. 8):

> A Nanomaterial is a material with any external dimension in the nanoscale [approximately 1 nm to 100 nm] and or having internal structure or surface structure in the nanoscale.

The European Commission's definition of a ***nanomaterial*** is as follows ("EU Nanomaterials Definition", 2017):

> A natural, incidental or manufactured material containing particles, in an unbound state or as an aggregate or as an agglomerate and where, for 50 % or more of the particles in the number size distribution, one or more external dimensions is in the size range 1 nm – 100 nm.
>
> In specific cases and where warranted by concerns for the environment, health, safety or competitiveness the number size distribution threshold of 50 % may be replaced by a threshold between 1 and 50 %.

In establishing the UDS, the rich array of actual and potential nanomaterials requires considerable detail to be differentiated from one another. It is extremely useful, however, to divide nanomaterials and the objects that contain them into four major types; each type of nanomaterial requires slightly different sets of information to describe it completely. The UDS differentiates between (1) an individual nano-object, (2a) a collection of identical nano-objects, (2b) a collection of different nano-objects, (3) a bulk material containing individually identifiable nano-objects, and (4) a bulk material that has nanoscale features.

It must be recognized that the distinction between different types of bulk materials may be difficult to determine, and the use of information categories related to those types depend on the application and discipline. At the same time, the functionality of nanomaterials may really take place as an individual nano-object or as a collection of a small number of nano-objects that have separated in use from the bulk material that originally contained it.

It should be noted that the applicability of the UDS is not limited to engineered or manufactured nanomaterials but is also pertinent to naturally occurring (processed or otherwise) nanomaterials. Because of the reactivity of nano-objects, the compositions, especially on the surface, can change significantly through its life cycle. This is discussed ahead.

Framework for characterizing nanomaterials

In the development of the UDS, input was drawn from a large variety of user communities, including standards committees such as ISO Technical Committee 229 Nanotechnologies (ISO TC 229, 2016) and ASTM Committee E56 on Nanotechnology ("ASTM E56 on Nanotechnology", 2017) as well as groups such as the OECD Working Party on Manufactured Nanomaterials (OECD WPMN, 2016). A framework of the information used by different disciplines in their nanomaterials work was created that integrated existing approaches that have focused on specific detailed aspects of nanomaterials, such as size, shape, and structure. The final framework defines four major information categories used to describe nanomaterials as shown in Table 9.3. Under each information category, there are numerous descriptors covering specific quantitative and qualitative data important in the description system.

Each of these information categories contains numerous subcategories that in turn contain the descriptors that provide the detailed data and information comprising a complete description system. The system is not hierarchical except that subcategories refer back to the main categories. Different users of the description system will use different subcategories and descriptors to a lesser or greater extent. These categories and subcategories can be used to create an ontology for nanomaterials that can be used to support many different types of applications.

The UDS identifies the various types of data and information, as shown in Figure 9.2, that can be used to describe a nanomaterial. It does not, however, prescribe which pieces of data and information *must* be reported; that is determined by the reason for describing a nanomaterial, which in turn is determined by the

Table 9.3 Major information categories used to describe a nanomaterial

Information category	Description
Characterization	A set of measurement results that taken together uniquely describes the physical, chemical, structural, and other characteristics of a nanomaterial
Production	A set of general and specific information that describes the production of a nanomaterial; the production of a nanomaterial is assumed to have a distinct initial phase followed by one or more post-production phases
Specification	A set of detailed information about specification documentation according to which a nanomaterial has been produced or documented
General identifiers	The general terms used to name and classify a nanomaterial

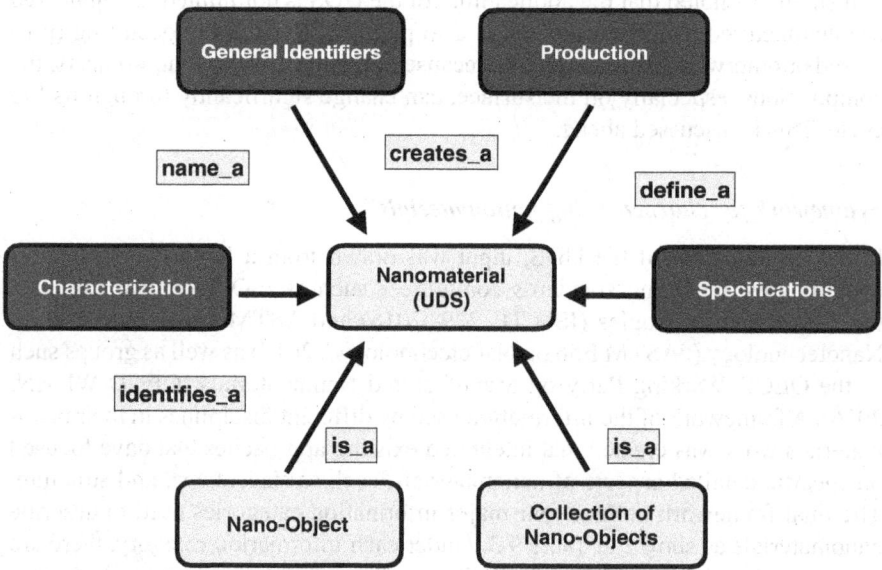

Figure 9.2 Framework for a Uniform Description System for nanomaterials

Note: Adapted from www.codata.org/nanomaterials.

community generating and using this data and information, that is, the context of the use. It should also be noted that additional descriptors may become necessary as our knowledge of the properties of nanomaterials increases.

Characterization of nano-objects

It is at the scale of individual nano-objects that the complexity and uniqueness of nanomaterials is most clearly demonstrated. The term nano-object is defined in

ISO TS 80004:1 (ISO TS, 2007) as *"a material with one, two or three external dimensions in the nanoscale"*. A nano-object is the smallest amount of a nanomaterial that is distinguishable. While normal uses of nanomaterials involve large numbers of nano-objects, it is important to be able to describe an individual nano-object accurately. The characterization of nano-objects comprises six major information categories as shown in Table 9.4.

Table 9.4 Information categories used to characterize nano-objects

Information category	Description
Shape	The most common criterion for defining the shape of a nano-object is its general three-dimensional geometry, or shape type. It is useful to define quantitative measures of shape, including *aspect ratio*, or sharpness, that reflects the fiber- or rod-like nature of a nano-object, *flatness*, or the lack of unevenness of a plate-like nano-object, and sphericity that provide an indication of how spherical is a nano-object.
Size	The very modifier "nano" illustrates the importance of size in describing a nano-object, yet size even on the nanoscale can vary greatly. The size of a cube-shaped nano-object with all three dimensions on the nanoscale can range from 1 nm^3 to 10^6 nm^3. Similarly, the surface area of a 100 nm-sided cube is 10^4 larger than a 1 nm-sided cube. Derived dimensions such as ballistic size are also reported. The dimensions needed to specify the size (internal and external dimensions) of different nano-objects vary according to the shape.
Chemical composition	The chemical make-up of a nano-object is a natural way to describe that object. Composition subtleties are complex and important. Nano-objects are very reactive and attract many different coatings, planned and random. Many nano-objects are inhomogeneous and have non-uniform chemical composition in their different parts. Many nano-objects are non-stoichiometric so that an accurate chemical composition is difficult to express.
Physical structure	A number of different physical structure models are possible, and their internal structures depend on their complexity. Some nano-objects are layered or shell-like, and others contain inhomogeneities. They can have features such as holes, protuberances, and appendages. Some nano-objects are synthesized to have specific pore sizes (e.g., for catalytic purposes).
Crystallographic structure	A nano-object can have multiple phases (physical structures) within it, each with a different crystallographic structure. The structure can be amorphous, polycrystalline, or monocrystalline.
Surface description	Structured surfaces on the nanoscale are produced to have unique and useful electronic and photonic properties. Because of the reactivity of its surfaces, a nano-object has adherents on its surface.

Characterization of collections of nano-objects

A collection of nano-objects is created either deliberately or through natural interactions and occurs during production, shipment, testing, and use. In most cases,
the reactivity of individual nano-objects means that on a practical scale, it is difficult to produce, manipulate, or use an individual nano-object in isolation of all
other nano-objects.

A collection is differentiated from bulk materials with nano-objects in that a
collection contains only nano-objects. There remains the ambiguity of an individual nano-object that has acquired adherents such as a full corona or partial
coverage. A collection of nano-objects may be homogeneous, composed of one
type of nano-object, or heterogeneous, composed on two or more different types.
Because of the wide diversity of possible collections, considerable thought must
be given to the details of accurately describing a collection. Collections of nano-
objects may vary in the number of nano-objects that are present. Examples of
collections include an aggregation or agglomeration of two or a small number of
similar or dissimilar nano-objects; a large number of similar or dissimilar nano-
objects attached to a substrate; a large number of (mostly) similar nano-objects
produced at the same time; a small or large number of nano-objects assembled for
shipping, transport, testing, or use; and a small or large number of nano-objects
assembled for inclusion in a larger amount of material. One of the major differen-
tiators between "small" and "large" collections is whether the resulting collection
can be treated as a nano-object itself. This is certainly the case when working with
agglomerates and aggregates involving a very small number (2–4) of particles.

Describing a collection can be done in diverse ways, and different types of
collections require different information for their description. In many situations, the description is made based on an *average* or *representative* collection.
The implications of this approach are significant. The correlation of properties
with specific collection features may be difficult. In a distribution of collections,
individual collections away from the *average* might exhibit significant levels
of reactivity and properties different from those that are *average*. Sampling of
a collection itself might change characteristics of the collection, adding to or
reducing, for example, the amount of association through a change of conditions.
Simple activities such as storage and transportation create dynamics that also
alter the characteristics of a collection. Table 9.5 summarizes the information
categories used to describe a collection of nano-objects.

Production of nanomaterials

For the purposes of the UDS, the production of a nanomaterial is assumed to
have a distinct initial production phase followed by one or more post-production
phases. The post-production phase may simply be storage after initial production
or a more complex transformation. ISO TC 229 has produced ISO 80004–8:2013,
which defines terminology applicable to nanomanufacturing ("ISO/TS 80004–
8:2013", 2017, pp. 80004–80008). In addition, much effort is being made by several engineering communities to develop process models that are applicable to a
wide variety of processes.

Table 9.5 Information categories used to characterize collections of nano-objects

General features	The description of a collection needs to include information on its general features (e.g., "typical", "proto-typical", "as sampled").
Composition	The composition of a collection of nano-objects is established by specifying the nature of each type of nano-object present and the amount thereof. The composition of larger collections can be problematic to determine.
Size distribution	Many production processes do not result in uniform nano-objects, especially with respect to size. The distribution of sizes of the nano-objects within a collection is an important determinant of its overall properties. Size distribution is a key element of the EU's definition of a nanomaterial and needs several pieces of data to report it correctly.
Physical structure	The physical structure of a collection of nano-objects is characterized by the arrangement of the individual nano-objects within it. A collection can have no structure (totally random), for example, if simply confined in a container or in a medium. A collection can have a regular or partially regular structure, for instance, if the collection is attached to a substrate of some type. Collections of nano-objects can have substructures if they are confined, such on a substrate.
Association	The most important classes of association used in describing collections of nano-objects are agglomeration and aggregation. These two classes are differentiated by the strength of the bonding holding the nano-objects together. ISO TC 229 has defined for these classes as follows (emphasis added).
	Agglomerate – Collection of *weakly bound* particles or aggregates or mixtures of the two where the resulting external surface area is similar to the sum of the surface areas of the individual components. The forces holding an agglomerate together are weak forces, for example van der Waals forces, or simple physical entanglement (ISO/TS 80004–2:2015, 2017).
	Aggregate – Particle comprising *strongly bonded or fused particles* where the resulting external surface area may be significantly smaller than the sum of calculated surface areas of the individual components. The forces holding an aggregate together are strong forces, for example covalent bonds, or those resulting from sintering or complex physical entanglement (ISO/TS 80004–2:2015, 2017).
Interfaces	An interface within a collection of nano-objects is defined as the boundary between two distinct spatial regions. An interface is described by its location, the two regions on either side of the boundaries, the boundary area, and the type and strength of the interaction.
Topology	Topology is the description of the overall connectivity and continuity of a collection of nano-objects or its components (where each component can be one or more nano-objects) or both, including the relative position in space of the components (e.g., totally or partially internal or external to each other, and their connectedness and boundaries). At present, there is no system under development to describe systematically the topological features of a collection of nano-objects.

The reactivity of individual and collections of nano-objects gives rise to questions about their stability under "non-reactive" conditions such as movement, temperature changes, exposure to heat, and shock (unexpected forces). The initial production includes information relevant as to how a nanomaterial was first synthesized, formulated, produced, or manufactured to achieve its primary structure and properties. The production of a nanomaterial in the context of a research or experimental environment is quite different from production in a commercial setting. The amount and type of the processing history information reported varies greatly depending on the circumstances as well as the source of the information. Many companies share very few processing details, relying instead on highlighting "unique" properties of their materials. Publicly funded research papers, however, might contain more complete details. Complete details of production within the UDS are available ("CODATA Nanomaterials WG", 2017).

Life cycle of nanomaterials

Because of their surface reactivity and quantum effects, nanomaterials can interact strongly with objects in their environment (Fadeel et al., 2015). While it is convenient to consider a nanomaterial as being created and then slightly altered through its life cycle (preparation for use, use, and disposal), the strong reactivity poses significant problems in describing a nanomaterial throughout its life cycle, which in turn creates problems for categorizing nanomaterials. Consider, for example, the life cycle associated with testing to determine properties. During the different steps in preparing and performing a test, a nanomaterial can be in at least four distinct states as shown in Table 9.6. A similar though slightly different life cycle exists for a nanomaterial being put into a product.

Table 9.6 States of a nanomaterial during the testing process

State	Description
As produced (manufactured, prepared, or natural nanomaterial)	This is the "substance" for which users, regulators, and the public want results; almost always a collection of nano-objects
As received	Despite precautions, changes occur during shipping and storage: agglomeration, aggregation, reactions, degradation
As prepared for testing	Usually some processing takes place, such as purification, dilution or concentration, reversal of shipping and storage effects
As sampled	A subset of the nanomaterial is taken for testing, using standard, organization-specific, or ad hoc procedures
In the test environment	Once in the test environment, the nanomaterial may experience reactions, additions, alterations, including coronas, surface modification, pH changes, etc. In addition, many characterization techniques are destructive; that is, there are techniques for which once the property of an individual nano-object is ascertained, the nano-object has been changed or no longer exists.

The question immediately arises about the meaning of measurement results and whether they should be assigned to the "original" nanomaterial "as produced", or to the modified nanomaterial in the test environment. For example, manufacturers often coat TiO_2 nano-objects with silica for better dispersability. Does this coating change the fundamental properties of the TiO_2 nano-objects such that categorization based on the "original" chemical composition is no longer valid?

Such considerations are not moot. As more nanomaterials are developed and pass into commerce, regulations and applications will be based on a variety of properties, and techniques such as read-across and QSAR will be used to predict the behavior of new and emerging nanomaterials (Gebel et al., 2014; Chen et al., 2016; Kuempel et al., 2012; Arts et al., 2015; Godwin et al., 2015; Oomen et al., 2015; Bos et al., 2015; Hristozov et al., 2016). The question arises to which state of the nanomaterials' life cycle should these predictive techniques be applied?

Use of the Uniform Description System for nanomaterials

In addition to supporting categorization, the UDS allows users, regardless of discipline, type of nanomaterial, or application, to use a common method for accurately describing a nanomaterial. Possible uses are outlined in Table 9.7.

Additional issues

Now that the UDS system has been motivated and sketched out here, we close with a few words on additional considerations that have influenced the development of the UDS system. First, recognize the relation between the UDS and categorization: one cannot categorize objects correctly if the features of the objects used to categorize them are ambiguously defined. QSAR technology was developed originally for chemical compounds that are well-defined and have structural features that can clearly be described. Identifying compounds with the same features is quite straightforward. Even if compounds are placed in a reactive environment, the compound "changes" in a well-defined way, for example, ionizes predictably. The concept of life cycle is not relevant.

Nanomaterials, because they are larger objects and highly reactive, change more dramatically, and the ability to define precisely what nano-object feature is causing a specific behavior is more difficult. This problem can be described more rigorously. The UDS provides a structure that allows detail specification of features (independent variables) that can be correlated with nanomaterial properties and functionalities. Other description systems, such as ISA-TAB-nano (Thomas et al., 2013) and the OECD test protocols (OECD WPMN, 2016), are incomplete and do not adequately detail all important characteristics of the tested object.

Next, it is worthwhile to consider the issues associated with using natural technical language for a description system and categorization rather than more formal modeling languages, ontologies, or even algorithmic schemes. The latter are designed to facilitate communication with and among computers. At present, however, natural technical language is primarily to facilitate communication among

Table 9.7 Uses of the Uniform Description System for nanomaterials

Nanoinformatics	Many groups are building data collections of measurement results, and users in turn want to use multiple data resources to gain access to all available information. The UDS provides a backbone for building the database schemas and ontologies that are at the core of nanoinformatics resources so that information from different resources can be integrated correctly.
Regulatory actions	The UDS provides a technology that allows regulators to define precisely and accurately specific nanomaterial(s) being regulated. General terms such as carbon nanotubes are not adequate for regulations. For example, certain forms of titanium oxide have toxic effects; other forms might not. The UDS allows specification of features that are to be regulated.
Standards developers	The UDS helps standards developers identify critical areas for standardization as well as the research needed to address those areas. For example, the description of the surface of a nano-object and the topology of a collection of nano-objects are areas in which no consensus approach yet exists to describe the complexity of nanomaterials.
Correlation of properties with nanomaterial features	The descriptors in the UDS can be considered as independent variables that affect in some way the properties of a nanomaterial. To be able to predict properties, one must identify and understand all the major variables that affect that property. The UDS provides a rigorous framework for systematically identifying and reporting the relationship between a feature (independent variable) and a property (dependent variable), which is of particular importance to health, safety, and environmental issues.
Researchers	As new nanomaterials are discovered and formed, an accurate description is necessary so that future researchers are able to perform studies on the same nanomaterial.
Purchase of nanomaterials	The complexity of nanomaterials precludes their specification by a simple name or formula. Purchasers of nanomaterials want to know exactly what they are getting, and providers of nanomaterials want to be able to clearly state what they are providing. The UDS provides a system to meet both needs.
Prediction of properties and evaluation of materials for use	The adoption of nanomaterials for use in products and other applications depends on the availability of reliable data about their performance under specified conditions. The UDS provides a mechanism for consistent reporting of data as well as the use of data from multiple sources in design and performance prediction software.

humans. It is likely that sometime in the future, description systems and categorization methods will primarily be computer tools, but that is not the case today. Given the range of people using both description systems and categorization methods, from lab researchers developing and testing new nanomaterials to nanoinformatics specialists to regulators to commercial producers, clarity of expression among humans is paramount. One of the problems in nano classification is determining what counts as a relevant property or structure, or a difference big enough to count as a different property or structure. Though resolving that vagueness might require human rather than computer analysis at this stage in nanomaterial research, the organization of property information must first be addressed. Let us look more closely at automated systems for both descriptions and categorization.

For description systems, first let us consider chemical compounds. After several decades of work, the major components of chemical composition have become almost fully automated. We can now easily identify atoms, bonds, spatial orientation, moieties, and chirality fairly precisely (Favre and Powell, 2013). Much of that knowledge is also expressible in machine-readable code such as InChI (Heller et al., 2015). Yet as described by other chapters in this volume, some chemical structure features still remain ambiguous when using these automated systems.

If we look at description systems for larger or more complex physical systems, such as bulk materials, ecological systems, and biological organisms, these are not yet well automated; that is, while database schemas and ontologies have been developed and information systems contain significant amounts of data, the generation of the composition and structure of such physical systems is not yet automated. The description of nanomaterials, though having requirements closer to those for chemical composition and structure than for bulk materials, and so forth, still are sufficiently complex that details and ambiguities cannot yet be handled solely by software. Until a better understanding of how to capture all such details is achieved, natural language should remain the choice for description systems.

The same situation exists for categorization with respect to automated processing of categorization criteria. For example, chemical structure databases can be searched and categorized with respect to chemical moieties (functional groups) and properties. This works because there are large comprehensive databases and software to decompose structure. For nanomaterials, we have neither large comprehensive databases nor software tools to process important information such as shape, size, chemical composition, and so forth. In time, this situation will improve. For example, there are efforts to develop models based on mathematical analysis of shapes and biological interactions, but this work is just beginning (Polo et al., 2016). Again, natural language categorization appears to be most appropriate.

Finally, one central question that needs to be asked is whether we can today identify successful categorization schemes for nanomaterials. Without going into what we mean by "successful", there are emerging schemes especially with respect to predicting environmental effects (e.g., Bos et al., 2015). As pointed out earlier, different contexts will support different categorization methods. Regardless, without large comprehensive databases of nanomaterials characterization and properties, the validity of any scheme will be in doubt.

Conclusions

One last set of thoughts is appropriate in this section, regarding practical and theoretical categorization approaches. The driving forces for workable categorization schemes of nanomaterials are very practical: nanomaterials are entering commerce, and many individuals and organizations have concerns about their long-term impact in both positive and negative ways. Can we understand and predict how nanomaterials provide important functionality such as drug delivery, drug efficacy, structural integrity, electronic and magnetic capabilities, and many more? Can we understand and predict when nanomaterials might present potential harm to human, other living things, and the environment?

In this chapter, we have tried to articulate some clear goals – scientific, methodological, and contextual – for categorization of nanomaterials. We have also summarized a system that provides a natural language way to describe nanomaterials, which is critical before any categorization scheme that is based on causality can be verified. We have also pointed out the necessity of large comprehensive databases of nanomaterials compositions, structures, and properties are needed before categorizations can be proved useful.

By proceeding systematically to address these concerns, progress will be made on new and accurate categorization schemes for nanomaterials. The Uniform Description System and its detailed information categories and descriptors provide as important new tool for more precise definitions of categorization schemes.

Acknowledgments

The author wishes to acknowledge support from the Future Nano Needs project supported by EU FP7 through contract 1234. Additional support, especially for the development of the UDS was provided by CODATA and the International Council for Science (ICSU). The author also wishes to thank Steve Freiman and Clayton Teague for many useful discussions.

References

Arts, J., Hadi, M., Irfan, M.-A., Keene, A., Kreiling, R., Lyon, D., Maier, M., Michel, K., Petry, T. & Sauer, U. (2015) A Decision-Making Framework for the Grouping and Testing of Nanomaterials (DF4nanoGrouping). *Regulatory Toxicology and Pharmacology.* 71(2), S1–S27.

ASTM E56 on Nanotechnology. (2017) Retrieved from www.astm.org/COMMITTEE/E56.htm 31 August 2018.

Bos, P., Gottardo, S., Scott-Fordsmand, J., van Tongeren, M., Semenzin, E., Fernandes, T., Hristozov, D., Hund-Rinke, K., Hunt, N. & Irfan, M.-A. (2015) The MARINA Risk Assessment Strategy: A Flexible Strategy for Efficient Information Collection and Risk Assessment of Nanomaterials. *International Journal of Environmental Research and Public Health.* 12(12), 15007–15021.

Bowker, G. & Star, S. (2000) *Sorting Things Out: Classification and Its Consequences.* Cambridge, MA: MIT Press.

Cambridge Crystallographic Data Centre. (2016) *Cambridge Crystallographic Data Centre.* Retrieved from www.ccdc.cam.ac.uk/.

Chen, G., Peijnenburg, W., Kovalishyn, V. & Vijver, M. (2016) Development of Nanostructure: Activity Relationships Assisting the Nanomaterial Hazard Categorization for Risk

Assessment and Regulatory Decision-Making. *Royal Society of Chemistry Advances.* 6(57), 52227–52235.

CODATA. (2017) *CODATA.* Retrieved from www.codata.org/ 31 August 2018.

CODATA Nanomaterials Working Group. (2017) *CODATA Nanomaterials.* Retrieved from www.codata.org/nanomaterials 31 August 2018.

EPA. (2018) *Chemical Substances When Manufactured or Processed as Nanoscale Materials: TSCA Reporting and Recordkeeping Requirements.* Retrieved from www.regulations.gov/document?D=EPA-HQ-OPPT-2010-0572-0137 7 January 2019.

EU Nanomaterials Definition. (2017) Retrieved from http://ec.europa.eu/environment/chemicals/nanotech/#definition 31 August 2018.

Fadeel, B., Fornara, A., Toprak, M. & Bhattacharya, K. (2015) Keeping It Real: The Importance of Material Characterization in Nanotoxicology. *Biochemical and Biophysical Research Communications.* 468(3), 498–503.

Favre, H. & Powell, W. (2013) *Nomenclature of Organic Chemistry: IUPAC Recommendations and Preferred Names 2013.* Cambridge: Royal Society of Chemistry.

FIZ Karlsruhe. (2016) Inorganic Crystal Structure Database. *Inorganic Crystal Structure Database.* Retrieved from www.fiz-karlsruhe.de/en/leistungen/kristallographie/icsd.html.

Fujita, T., Iwasa, J. & Hansch, C. (1964) A New Substituent Constant, Π, Derived from Partition Coefficients. *Journal of the American Chemical Society.* 86(23), 5175–5180.

Gebel, T., Foth, H., Damm, G., Freyberger, A., Kramer, P.-J., Lilienblum, W., Röhl, C., Schupp, T., Weiss, C. & Wollin, K.-M. (2014) Manufactured Nanomaterials: Categorization and Approaches to Hazard Assessment. *Archives of Toxicology.* 88(12), 2191–2211.

Godwin, H., Nameth, C., Avery, D., Bergeson, L., Bernard, D., Beryt, E., Boyes, W., Brown, S., Clippinger, A. & Cohen, Y. (2015) Nanomaterial Categorization for Assessing Risk Potential to Facilitate Regulatory Decision-Making. *ACS Nano.* 9(4), 3409–3417. DOI: 10.1021/acsnano.5b00941.

Gramatica, P. (2007) Principles of QSAR Models Validation: Internal and External. *Molecular Informatics.* 26(5), 694–701.

Hansch, C. & Fujita, T. (1964) P-σ-π Analysis: A Method for the Correlation of Biological Activity and Chemical Structure. *Journal of the American Chemical Society.* 86(8), 1616–1626.

Hansen, S., Jensen, K. & Baun, A. (2014) NanoRiskCat: A Conceptual Tool for Categorization and Communication of Exposure Potentials and Hazards of Nanomaterials in Consumer Products. *Journal of Nanoparticle Research.* 16(1), 2195.

Heller, S., McNaught, A., Pletnev, I., Stein, S. & Tchekhovskoi, D. (2015) InChI, the IUPAC International Chemical Identifier. *Journal of Cheminformatics.* 7(1), 23.

Hill, A. B. (1965) Association or Causation? *Proceedings of the Royal Society of Medicine.* 58, 295–300.

Hristozov, D., Gottardo, S., Semenzin, E., Oomen, A., Bos, P., Peijnenburg, W., van Tongeren, M., Nowack, B., Hunt, N. & Brunelli, A. (2016) Frameworks and Tools for Risk Assessment of Manufactured Nanomaterials. *Environment International.* 95, 36–53.

Isaacson, Walter. (2014) *The Innovators: How a Group of Inventors, Hackers, Geniuses and Geeks Created the Digital Revolution.* New York: Simon and Schuster.

ISO TC 229. (2016) ISO Technical Committee 229 Nanotechnologies. *ISO Technical Committee 229 Nanotechnologies.* Retrieved from www.iso.org/iso/iso_technical_committee?commid=381983.

ISO TS. (2007) *80004–1: Nanotechnologies-Vocabulary-Part 1: Core Terms.* Geneva, Switzerland: International Standards Organization.

ISO/TS 80004–2:2015. (2017) Retrieved from www.iso.org/standard/54440.html 31 August 2018.

ISO/TS 80004–8:2013. (2017) Retrieved from www.iso.org/standard/52937.html 31 August 2018.

Karcher, S., Willighagen, E., Rumble, J., Ehrhart, F., Evelo, C. T., Fritts, M., Gaheen, S., Harper, S. L., Hoover, M. D., Jeliazkova, N. & Lewinski, N. (2018) Integration among Databases and Data Sets to Support Productive Nanotechnology: Challenges and Recommendations. *NanoImpact*. 9, 85–101.

Kuempel, E., Geraci, C. & Schulte, P. (2012) Risk Assessment and Risk Management of Nanomaterials in the Workplace: Translating Research to Practice. *Annals of Occupational Hygiene*. 56(5), 491–505.

Mannhold, R., Krogsgaard-Larsen, P. & Timmerman, H. (2008) *QSAR: Hansch Analysis and Related Approaches*. Vol. 1. Hoboken, NJ: John Wiley & Sons.

Martens, H. & Martens, M. (2001) *Multivariate Analysis of Quality: An Introduction*. Chichester: Wiley.

Nel, A., Xia, T., Meng, H., Wang, X., Lin, S., Ji, Z. & Zhang, H. (2012) Nanomaterial Toxicity Testing in the 21st Century: Use of a Predictive Toxicological Approach and High-Throughput Screening. *Accounts of Chemical Research*. 46(3), 607–621.

NIH Pub Chem. (2018) *The PubChem Project*. Retrieved from https://pubchem.ncbi.nlm.nih.gov/ 7 January 2019.

Oomen, A., Bleeker, E., Bos, P., van Broekhuizen, F., Gottardo, S., Groenewold, M., Hristozov, D., Hund-Rinke, K., Irfan, M.-A. & Marcomini, A. (2015) Grouping and Read-Across Approaches for Risk Assessment of Nanomaterials. *International Journal of Environmental Research and Public Health*. 12(10), 13415–13434.

Organisation for Economic Co-Operation and Development. (2016) OECD Working Party on Manufactured Nanomaterials. *Safety of Nanomaterials*. Retrieved from www.oecd.org/science/nanosafety/ 17 February 2019.

Phillips, C. & Goodman, K. (2004) The Missed Lessons of Sir Austin Bradford Hill. *Epidemiologic Perspectives & Innovations*. 1(1), 3.

Polo, E., Castagnola, V. & Dawson, K. (2016) Understanding and Characterizing Functional Properties of Nanoparticles. In *Pharmaceutical Nanotechnology: Innovation and Production*. 2 vols. Hoboken, NJ: Wiley.

Roco, M. (2011) *The Long View of Nanotechnology Development: The National Nanotechnology Initiative at 10 Years*. Berlin: Springer.

Rumble, J. (2018) *CRC Handbook of Chemistry and Physics 98th Edition*. Retrieved from http://hbcponline.com/faces/contents/ContentsSearch.xhtml;jsessionid=3C8968118A87C0DE65E568F9C8F3EEEB 7 January 2017.

Rumble, J. & Freiman, S. (2012) Describing Nanomaterials: Meeting the Needs of Diverse Data Communities. *Data Science Journal*. 11, ASMD1–ASMD6.

Rumble, J., Freiman, S. & Teague, C. (2014) The Description of Nanomaterials: A Multi-Disciplinary Uniform Description System. 2014 IEEE International Conference Bioinformatics and Biomedicine (pp. 34–39).

Sellers, K., Mackay, C., Bergeson, L., Clough, S., Hoyt, M., Chen, J., Henry, K. & Hamblen, J. (2008) *Nanotechnology and the Environment*. Boca Raton, FL: CRC Press.

Thomas, C., George, S., Horst, A., Ji, Z., Miller, R., Peralta-Videa, J., Xia, T., Pokhrel, S., Mädler, L. & Gardea-Torresdey, J. (2011) Nanomaterials in the Environment: From Materials to High-Throughput Screening to Organisms. *ACS Nano*. 5(1), 13–20.

Thomas, D., Gaheen, S., Harper, S., Fritts, M., Klaessig, F., Hahn-Dantona, E., Paik, D., Pan, S., Stafford, G. A., Freund, E. T. & Klemm, J. D. (2013) ISA-TAB-Nano: A Specification for Sharing Nanomaterial Research Data in Spreadsheet-Based Format. *BMC Biotechnology*. 13(1), 1.

Vance, M., Kuiken, T., Vejerano, E., McGinnis, S., Hochella, M., Rejeski, D. & Hull, M. (2015) Nanotechnology in the Real World: Redeveloping the Nanomaterial Consumer Products Inventory. *Beilstein Journal of Nanotechnology*. 6, 1769.

Versailles Project on Advanced Materials and Standards. (2017) *VAMAS*. Retrieved from www.vamas.org/ 31 August.

Woodward, J. (2003) *Making Things Happen: A Theory of Causal Explanation*. Oxford: Oxford University Press.

Xia, Y. (2014) Are We Entering the Nano Era? *Angewandte Chemie International Edition*. 53(46), 12268–12271.

Index

Note: page numbers in *italic* indicate a figure and page numbers in **bold** indicate a table on the corresponding page.